T0223939

The ORIGIN of STARS

The ORIGIN of STARS

Michael D Smith

Imperial College Press

ICP

Published by

Imperial College Press
57 Shelton Street
Covent Garden
London WC2H 9HE

Distributed by

World Scientific Publishing Co. Pte. Ltd.
5 Toh Tuck Link, Singapore 596224
USA office: 27 Warren Street, Suite 401-402, Hackensack, NJ 07601
UK office: 57 Shelton Street, Covent Garden, London WC2H 9HE

British Library Cataloguing-in-Publication Data
A catalogue record for this book is available from the British Library.

THE ORIGIN OF STARS

ISBN-13 978-1-86094-489-5
ISBN-10 1-86094-489-2

ISBN-13 978-1-86094-501-4 (pbk)
ISBN-10 1-86094-501-5 (pbk)

Printed in Singapore

To Daniela Ellen

Preface

It may be folly to take time out to write a book during the turbulence of a revolution. Many basic facts in the field of star formation are not being established but being ousted. Standard scenarios which were widely accepted are now being re-worked into a fresh emerging picture.

The purpose of this book is to describe the methods, results and theory surrounding the birth of stars so as to make the entire subject accessible within a single volume. I give high priority to ease of access to all interested in understanding a vital part of our origins. The hope is that the fascinating story of star birth, with the contentious issues and the conflicting interpretations, will hold the attention of the reader to the finish. Thus, the book is meant to be self-contained and, as far as possible, self-explanatory.

It may be wise to record the changes made up to this moment and to stimulate the directions for further progress. The field has become popular for research because we have found techniques to penetrate the birth sites. It is still only just possible to cover the entire subject in one volume, placing all the observations and models within a broad and unifying overview without, hopefully, a biased or over-convinced approach.

My wish is to make it possible for the average science reader to understand how stars are conceived, born and grow towards adulthood. The book can be read as an introduction or as a guide but not as a complete reference manual. A modest amount of mathematics cannot be avoided although, through quantification and explanation around the equations, I aim to permit the determined reader to reach a level from which new data and new interpretations can be confidently evaluated.

This book represents the work of an entire community of star formation researchers. I have merely tried to accumulate, communicate and explain. It would be unfair to credit the contributions of specific individuals. However, I would like to thank my close colleagues Mark Bailey, Chris Davis, Mordecai Mac Low and Hans Zinnecker for their encouragement and Peter Brand, Dirk ter Haar, Colin Norman, Harold Yorke and the late Dennis Sciama for their guidance over the years.

Above all, I cannot praise enough my wife Daniela for insight and inspiration. Daniela is with me in everything I do.

Michael D. Smith

Contents

Chapter 1

Introduction

1.1 Our Perception

Fifty years ago, the suspicion was raised that star birth was a thing of the past. The presence of young stars in an old galaxy seemed impossible. Even recently, little was known about how stars presently form and the birth itself remained an absolute mystery. There was a mute gap left exclusively to hand-waving gestures. Even in the best circumstances, we were restricted to weak evidence which often led to indecisive or contradictory conclusions. Today, we find ourselves immersed in facts as the nature of the birth emerges. The revolution was instigated by technological advances which have enabled us to make the necessary observations. Centuries of error-prone and conflicting thoughts can now be laid aside. Stars are indeed being born profusely even in our own back yard.

Stars have a dramatic life history. They are conceived within obscuring maternal clouds and are born through rapid contractions and violent ejections. They grow with others with whom they interact. And they often die in isolation having lived a life unique in some way. Yet, stars are inanimate objects, worthy of scientific endeavour and, apart from the Sun, of no direct bearing on the human condition.

The human condition, however, improves through revelation. Insight and discovery evoke profound feelings of excitement and inspiration which help raise the quality of our lives. Although we live in awe of the stars and we depend upon them for our existence, the mystery of their origin has been maintained. This has been in spite of our deep understanding of the intricacies of how they subsequently evolve and die. Yet the death of stars proves to be important to the conception of life, both stellar and human. This provokes the Genesis question: what made the first star in the Universe?

The quest for the 'holy grail' of star formation was the quest to detect the moment of birth. The decisive journey was along an evolutionary track which connects a cloud to a star. The challenge was to complete the journey by finding the connection between the collapsing cloud and the emergent young star. To do so, we developed telescopes, cameras and receivers which could probe deep into the clouds. In optical light, the clouds are opaque. In light with longer wavelengths – infrared and millimetre radiation – the clouds become transparent. For the first time, our view penetrates the clouds and we can discuss how stars form.

The tremendous advances in our knowledge make this book almost unrecognisable from previous books on the subject. There have been five other major upheavals in the subject in the last few years. They also make for a vast quantity of fascinating material, none of which should be excluded. The story has reached epic proportions.

1.2   The Story of Star Formation

The significance of the six upheavals can be realised by placing them into the story of formation. The chapters are thus ordered to follow the storyline which leads to a star like our Sun, as sketched in Fig. 1.1.

We first introduce the rudimentary materials from which stars are made. Stars are seen to be forming in giant clouds of molecules within the tenuous interstellar medium (Stage 1 of Fig. 1.1). It has long been intriguing that the interstellar gas has not been exhausted and that star formation continues despite the advanced age of our Galaxy.

The basic tools that we possess to work the materials are a range of interacting physical, chemical and radiation processes (§1.5). This provides an essential background that permits observations of the interstellar medium and molecular clouds to be placed into perspective (§2.5).

This leads to the classical theory of star formation in which the opposing forces of gravity and pressure provide a precarious equilibrium. The gradual collapse through states of hydrostatic equilibrium is laboured but was consistent with the fact that star formation continues in our Galaxy (§3.7).

However, the slow evolution and delicate balance contradict almost all other observational evidence as well as our computer simulations. We find that individual clouds are simply too young and too turbulent. We ob-

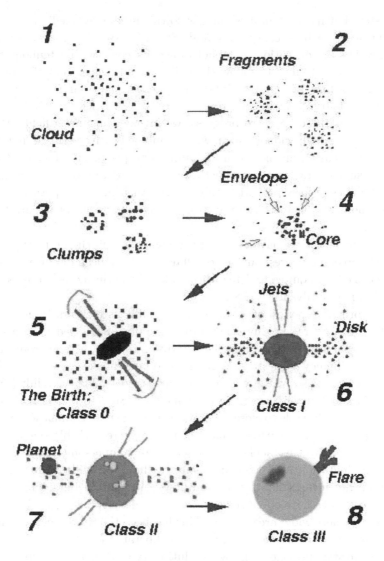

Fig. 1.1   A sketched guide to the stages in the formation of a low mass star like our Sun from conception through early development.

viously need a hydrodynamic approach to describe the fragmentation into clumps (Stages 2 & 3 of Fig. 1.1). After 50 years of stagnation, the concepts of turbulence are no longer in chaos and in §4.4 we see how a turbulent revolution is solving long-standing problems. We now interpret the sustained star formation as one of the effects of turbulence: stirring the gas generates some stars but recycles most of the gas to the interstellar medium.

Out of the turbulence, eggs are laid (Stage 4). We have now uncovered the sites of star conception: compact molecular cores within tenuous surroundings. We attempt to follow the collapse and contraction of cores as a pure gas or fluid flow and find the way blocked: the spinning gas will be held up by its own centrifugal force.

To permit the collapse to progress, we invoke a magnetic field. Magnetic flux threads the clouds which, when combined with the core dynamics, leads to fascinating magnetohydrodynamic concepts (§6.6). The gas thermal pressure interacts with magnetic pressure and tension. In addition, magnetic lines of force are twisted by the rotating fluid and, perhaps later or elsewhere, uncoil to rotate the fluid. The collapse problems are then solvable.

We now believe we have detected the young star in the nest (Stage 5). Not only pre-stellar cores, but protostellar cores have been discovered. The mammoth effort required to detect these cold contracting cores is the second of the six major advances (§7.8).

The formative period of a young star is millions of years (Stages 6 & 7). The baby stars are tiny centres of attraction, being fed from a disk supplied from an envelope. The disk continues to feed the star from birth through to infancy. Questions remain as to what makes the disk such a reliable source. Finally, the young star starts to resemble the adult (main sequence) star but with exaggerated behaviour (Stage 8). It is no longer nurtured but still liable to close encounters which can be traumatic (§8.10).

Since the birth, a protostar has been ejecting remarkable slender jets of gas which drive spectacular outflows to highly supersonic speeds. This yields the inflow-outflow enigma: when searching for evidence for inflow, we often find just outflow. Many outflows turn out to be gigantic twin structures, which have only been detected after developing innovative technology to explore wider areas of sky. This has been the third great advance, the consequences of which are still being deliberated (§9.8).

Massive stars provide their own evolutionary problems. They live short and intense lives during which they have a disproportionate influence on events around them. Their positive feedback is the triggering of new generations of stars through compression; the negative feedback is the disruption of potential cores in their vicinity. As a result, we suspect that much of star formation is *self-regulated* (§10.10).

Stars are not isolated. They tend to form as members of binaries or small systems. The systems form and evolve almost exclusively in clusters.

The gas supply to the individual depends on their location in the cluster and the competition within the cluster and within the system. Observation and theory then demonstrate principles of mass segregation. These studies of the dynamical evolution, with remarkable spatial resolution provided by the latest generation of telescopes and supercomputer simulations, comprise the fourth great advance (§11.9).

The amazing discovery of very low-mass objects in these same regions of star formation is the fifth advance. The objects which will never become stars are brown dwarfs and free-floating objects with the sizes of planets. Their origins and relationship to the wide range of initial masses are crucial tests for all our ideas. Are they failed stars, ejected members or independent characters? In addition, planet formation can only be studied as part of the star formation process. The planet Earth was constructed during the formation of the Sun, probably during the late Class II phase when the disk was still quite active (§11.9).

Where and when appeared the very first star? In the early Universe, only basic commodities were present. The first star in the Universe preceded the first supernova. This is obvious but, in the absence of supernova enrichment, primordial stars must have formed in *warm atomic* gas. What this means and how this has determined the character of our present day Universe is just coming to light. The latest satellite observations and highly sophisticated simulations indicate that the primordial stars appeared first and extremely early. This, the sixth great advance, heads our final chapter which investigates the galactic scale, the super star cluster and the starburst galaxies (§12.8).

In terms of physics and chemistry, the story is too complex. However, scientists and mathematicians are developing tools to describe the *dynamics* of complex systems. In the future, we can combine principles of turbulence, self-regulation, self-propagation and self-destruction to construct working models.

1.3 The Early History

There is little early history surrounding the general subject of Star Formation. All the attention has been focused upon the origin of a single stellar system. As outlined below, some renowned individuals have contemplated the origin and early development of the solar system. Many of the ideas will resurface in modern theories.

René Descartes proposed a Theory of Vortices in 1644. He postulated that space was entirely filled with swirling gas in various states. The friction between the vortices 'filed' matter down and funnelled it towards the eye of the vortex where it collected to form the Sun. Fine material formed the heavens on being expelled from the vortex while heavy material was trapped in the vortex. Secondary vortices around the planets formed the systems of satellites.

Emanuel Swedenborg put forward a Nebula Hypothesis in 1734. The Sun was formed out of a rapidly rotating nebula. The planets were the result of condensations from a gauze ejected out of the Sun. The germinal idea for his nebular hypothesis came from a seance with inhabitants of Jupiter.

Georges Buffon suggested an Impact Theory in 1745. He proposed that a passing comet grazed the Sun and tore some material out of it. This cooled and formed the Earth. Apparently, Buffon had in mind a comet as massive as the Sun and an encounter corresponding to a modern tidal theory.

Immanuel Kant (1755) and Pierre Simon de Laplace (1796) independently proposed Nebular Hypotheses, amongst the oldest surviving scientific hypotheses. They involved a rotating cloud of matter cooling and contracting under its own gravitation. This cloud then flattens into a revolving disk which, in order to conserve angular momentum, spins up until it sheds its outer edge leaving successive rings of matter as it contracts. Kant tried to start from matter at rest whereas Laplace started with an extended mass already rotating.

Charles Messier (1771) recorded the shapes of astrophysical nebulae which were suggestive of disks around stars in which new planets might be forming. Even though these nebulae turned out to be galaxies, the Kant-Laplace hypothesis still survives.

George Darwin, son of Charles Darwin, conjured up a Tidal Hypothesis in 1898. Extrapolating back in time, to four million years ago, the moon was pressed nearly against the Earth. Then, one day, a heavy tide occurred in the oceans which lifted the moon out.

Sir James Jeans investigated Gravitational Instability in the early 20th century. He demonstrated that there was a minimum amount of gas that will collapse under its own self-gravity. Jeans calculated that above this critical value, gravity would overcome the thermal motion of the particles which would otherwise disperse the cloud. The critical mass of gas is called the Jeans Mass.

Thomas Chamberlin (1901) and Forest Moulton (1905) proposed a planetesimal hypothesis. They postulated that the materials now composing the Sun, planets, and satellites, at one time existed as a spiral nebula, or as a great spiral swarm of discrete particles. Each particle was in elliptic motion about the central nucleus. James Jeans (1916) and Harold Jeffreys proposed a new Tidal Hypothesis in 1917 while World War I was in progress. A passing or grazing star is supposed to have pulled out a long cigar-shaped strand of material from the Sun. The cigar would fragment to form the planets with the larger planets at intermediate distances.

1.4 Modern History

In the 1930s, it was realised that stars are powered through most of their lives by thermonuclear reactions which convert hydrogen to helium. With the lack of observations, however, the suggested models could only be tested against physical principles. For example, Lyman Spitzer's 1939 refutation of tidal theory brought down the Jeans-Jeffreys' hypothesis. He showed that the material torn out of the Sun by near-collisions would be hot and so would then rapidly expand and never contract into planets. Nevertheless, ideas were plentiful.

Henry Russell's Binary and Triple Star Theories (1941) resemble Buffon's passing star theory. The Sun was originally part of a binary system and the second star of this system then underwent a very close encounter with a third star. The encounter ejected a gaseous filament in which the planets formed.

Raymond Lyttleton proposed the Triple Star Theory in 1941 in which the Sun was originally part of a triple star system. The Sun's companions accreted gas and grew closer and closer together until they fragmented because of rotational instabilities. After merging, the stars form planets.

Fred Hoyle put forward a Supernova Hypothesis in 1944. Hoyle, inspired by Lyttleton's system, set up a system in which the Sun's companion star was a supernova. The outburst would have to be sufficient to break up the binary system yet leave sufficient remains to form the planets.

Fred Whipple promoted the Dust Cloud Hypothesis in 1946, applicable to the origin of all stars. The pressure of light rays from stars pushed dust into clouds, and chance concentrations condensed into stars. A smaller dust cloud was then captured with a much greater angular momentum, enough to form the planets. Whipple had thus proposed a mechanism to trigger stars.

Carl von Weizsäcker revived the Nebula Hypothesis in 1944. An extended envelope surrounds the forming Sun. Large regular turbulent eddies form in a disk containing one tenth of a solar mass. He realised the significance of supersonic motion and magnetic coupling of the dust to the gas.

Dirk ter Haar (1950) discarded the large regular vortices and found that gravitational instabilities would also be ineffective in the thick solar nebula. He thus proposed collisional accretion into condensations. The problem he raised, however, was that the turbulence would decay before sufficient collisions would build up the condensations. The turbulence would have to be driven but there was no apparent driver. This problem was to return again in the 1990s but on a much larger scale.

Von Weizsäcker then put forward a Rejuvenation Hypothesis in 1951. The existence of young stars within an aged Galaxy is a contradiction. Turbulence and gravity imply a formation time of 5 million years, a thousand times shorter than the age of the Galaxy. Therefore, all the gas should have been consumed. It would take over forty years to solve this dilemma. In the meantime, von Weizsäcker was led to the 'suspicion' that star formation was impossible! Instead, supposedly young stars were old stars, being rejuvenated by accumulating gas which lay in their paths.

1.5　Summary

The star formation history serves as a warning to ignore speculation in this book. The theories have flourished on being qualitative and floundered when subject to quantitative analysis. To try to avoid repeating history, we will not shy away from a quantitative approach where appropriate. To the credit of their founders, however, we will meet again many of the above concepts in revamped forms.

Chapter 2

The Physics and Chemistry

To appreciate the following story, we need to be familiar with the setting, the cast, their observational appearance and their physical interactions. Here, each member of the cast – gas, dust, cosmic rays, magnetic fields and radiation – is introduced and their potential behaviour is studied.

Stars form in clouds. Like atmospheric clouds, they are largely molecular and opaque. However, star-forming clouds are made more complex by a quite stunning range of physical, chemical and dynamical processes. The processes work in synthesis to provide triggers, regulators and blockers during the collapse stages. Energy is transferred from large scales down to small scales. Simultaneously, feedback occurs from small scales back onto large scales. As we shall see, the progress of a parcel of gas is far from systematic – at each stage in the development there is a chance that it will be rejected. It will, therefore, prove rewarding to first acquire a broad overview.

2.1 Scales and Ranges

At first sight, when interstellar clouds and atmospheric clouds are compared, they appear to have little in common. According to Table 2.1, they could hardly be more different. Yet they do resemble each other in other respects. Besides both being molecular and opaque, they both produce magnificent reflection nebula when illuminated by a nearby star, as illustrated in Fig. 2.1. Most relevant, however, is that they are both ephemeral: they are transient with lifetimes as short as their dynamical times (as given by their size divided by their typical internal speed, as listed in Table 2.1). In other words, their own swirling, turbulent motions disperse the clouds rapidly.

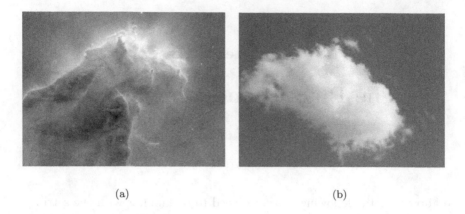

(a)   (b)

Fig. 2.1   Molecular clouds and atmospheric clouds. Similar hydrodynamic processes shape (a) interstellar clouds and (b) cumulus clouds in the sky despite contrasting scales. Both are illuminated by stars. Image (a) is a detail from the Eagle Nebula (M16) displayed fully in Fig. 2.2 (Credit: (a) J. Hester & P. Scowen (Arizona State U.), HST & NASA) and (b) Shawn Wall.

Table 2.1   A comparison of scales between typical molecular and atmospheric clouds.

	Molecular Cloud	Atmospheric Cloud
Size	10^{14} km	1 km
Mass	10^{36} gm	10^{11} gm
Particle density	10^3 cm^{-3}	10^{19} cm^{-3}
Temperature	20 K	260 K
Mol./atomic weight	2.3	29
Speed of sound	0.3 km s^{-1}	0.3 km s^{-1}
Turbulent speed	3 km s^{-1}	0.003 km s^{-1}
Dynamical time	Million years	Five minutes

Why hasn't there been a consensus on the star formation mechanism, given that there are only a few components to interact? The answer lies in the range of physical, chemical and dynamical processes which link them. This results in a variation of the dominant laws from stage to stage. In addition, we meet extreme regimes way beyond our experience, as summarised in Table 2.2.

High compressibility provides the major conceptual distinction of an interstellar cloud from a cumulus cloud. A parcel of gas from the hot interstellar medium in our Galaxy would decrease its volume by a factor of 10^{26} on its way to becoming a star, passing through the states shown in Table 2.2. That is, its number density increases from $0.01\,\mathrm{cm}^{-3}$ to $10^{24}\,\mathrm{cm}^{-3}$ (compare to the molecular density of air of $10^{19}\,\mathrm{cm}^{-3}$), and so reaches a mass density of almost $1\,\mathrm{g\,cm}^{-3}$.

We will encounter interstellar gas with a wide range in temperatures. Molecular clouds are cold (5–50 K) but protostars send shock waves into the clouds, capable of temporarily raising the temperature to over 10,000 K without destroying the molecules (§10.5). Atoms can be raised to temperatures in excess of a million Kelvin through faster shock waves and in the highly active coronae of young stars (§9.7).

Table 2.2   Major star formation scales. The temperature, T, is in Kelvin and the final column lists the dynamical time scale in seconds.

Phase	Size (cm)	Density $\mathrm{g\,cm}^{-3}$	T (K)	Time (s)
Atomic ISM	10^{21}–10^{20}	10^{-26}–10^{-22}	10^{6}–10^{2}	10^{15}
Molecular cloud	10^{20}–10^{18}	10^{-22}–10^{-18}	10^{2}–10^{1}	10^{14}
Protostar collapse	10^{18}–10^{12}	10^{-18}–10^{-3}	10^{1}–10^{6}	10^{13}
Pre-main-seq. contraction	10^{12}–10^{11}	10^{-3}–10^{0}	10^{6}–10^{7}	10^{15}

Lengths are measured in three units according to how best to appreciate the scale in discussion. Thus, cloud sizes are measured in terms of parsecs ($1\,\mathrm{pc} = 3.09 \times 10^{18}\,\mathrm{cm}$), cores, stellar and planetary systems in terms of Astronomical Units ($1\,\mathrm{AU} = 1.50 \times 10^{13}\,\mathrm{cm}$) and single stars in terms of the solar radius ($1\,\mathrm{R_{\odot}} = 6.96 \times 10^{10}\,\mathrm{cm}$). Hot diffuse gas occupies galactic kiloparsec scales while gas accretes to within a few solar radii of the growing young star. This is a difference in scale of ten magnitudes of ten from $10^{21}\,\mathrm{cm}$ to $10^{11}\,\mathrm{cm}$.

Time is usually given in years, that is 3.15×10^{7} seconds. Speeds, however, are expressed in centimetres or kilometres per second, whichever is appropriate for the phenomena studied.

Mass is usually expressed in solar units rather than grams ($1\,\mathrm{M_{\odot}} = 1.99 \times 10^{33}\,\mathrm{g}$) and gravitational acceleration is given by GM/R^{2}

for a distance R from a mass M, where $G = 6.67 \times 10^{-8} \, cm^3 \, g^{-1} \, s^{-2}$ is the gravitational constant.

While the mass density of gas is given in $gm \, cm^{-3}$, it is often useful to measure the mass per cubic parsec for the stellar content where $1 \, M_\odot \, pc^{-3} = 6.8 \times 10^{-23} \, g \, cm^{-3}$. A particle density is often more appropriate but can often be confusing since we employ the hydrogen atomic density, $n(H)$, hydrogen molecular density $n(H_2)$, hydrogen nucleon density $(n = n(H) + 2n(H_2))$, and the total particle density, n_p including an extra $\sim 10\%$ of helium atoms (above the hydrogen nucleonic number) which contribute 40% more by mass. The abundances of other elements are small and can be neglected in the overall mass budget of a cloud.

Luminosity and power are expressed in solar units $(1 \, L_\odot = 3.83 \times 10^{33} \, erg \, s^{-1})$ where the erg is the CGS abbreviation for $1 \, g \, cm^2 \, s^{-2}$ of energy. The KMS unit of Watt $(1 \, W_\odot = 10^7 \, erg \, s^{-1})$ is also often employed. Observers find that the traditional 'magnitude' unit provides a convenient logarithmic scaling of luminosity. The magnitude system is particularly convenient when extinction is in discussion since it is linearly related to the amount of obscuring material, whereas the luminosity falls off exponentially. Magnitudes will be defined and employed in §2.4.1.

Expressions for energy and temperature present the greatest variety. Astrochemists and X-ray astronomers often discuss in terms of electron Volts where $1 \, eV = 1.60 \times 10^{-12} \, erg$ (see Table 2.3). The frequency of radiation, ν, also converts to an energy $h\nu$ where h is the Planck constant, $h = 6.63 \times 10^{-27} \, erg \, s$. As an example, ionisation of cold H_2 requires an energy exceeding $15.4 \, eV$. So, photons with $\nu > 3.72 \times 10^{15}$ Hz are required which, according to Table 2.3, are ultraviolet photons.

The excitation energies of atoms and molecules are often expressed as a temperature using $E = kT$ where $k = 1.38 \times 10^{-16} \, erg \, K^{-1}$ is the Boltzmann constant.

Wavelengths are often more convenient than frequencies since the length of a wave can be directly compared to the size of atoms, molecules and dust particles. Thus, the Ångstrom, $1 \, \text{Å} = 10^{-8} \, cm$, and the micron, $1 \, \mu m = 10^{-4} \, cm$, are often useful. Units for radiation and wavelengths are summarised for reference in Table 2.3.

Table 2.3 Wavelengths and energies relevant to star formation studies.

Regime	Wavelength	Energy	Frequency (Hz)
Radio/millimetre	0.1–1000 cm		3×10^{7}–3×10^{11}
Sub-millimetre	300–1000 μm		3×10^{11}–1×10^{12}
Far-infrared	10–300 μm		1×10^{12}–3×10^{13}
Near-infrared	0.8–10 μm		3×10^{13}–4×10^{14}
Optical	4000–8000 Å		4×10^{14}–7×10^{14}
Ultraviolet	3000–4000 Å		7×10^{14}–1×10^{15}
Far UV	912–3000 Å	4–13.6 eV	1×10^{15}–3×10^{15}
Extreme UV	100–912 Å	13.6–100 eV	3×10^{15}–2×10^{16}
Soft X-ray	-	0.1–2 keV	2×10^{16}–4×10^{17}
X-ray	-	2–1000 keV	4×10^{17}–2×10^{20}
Gamma ray	-	1–1000 MeV	2×10^{20}–2×10^{23}

2.2 The Ingredients

2.2.1 *Atoms, molecules and dust*

The interstellar medium, or ISM, is a broad name for all that exists between the stars within galaxies. It includes diverse clouds of gas, which are composed mainly of atoms, molecules and ions of hydrogen, (H), electrons (e), as well as small amounts of other heavier elements in atomic and molecular form. There is a constant composition by number of about 90% hydrogen and 9% helium. The abundances of other atoms, mainly carbon, oxygen or nitrogen, vary depending on the enrichment due to stellar nucleosynthesis (the creation of new atoms by fusion in stars) and the removal due to condensation onto solid particles called dust grains (see below).

Molecules have been found concentrated in dense aggregates called molecular clouds. These are cold and dark regions in which hydrogen molecules outnumber other molecules by 1000 to 1 on average. This remains true until the heavy elements held in dust grains (see below) drift to the midplane of a disk surrounding a young star, the location where planets may eventually form.

Before 1970 there was little evidence for interstellar molecules. This all changed when millimetre, infrared and ultraviolet astronomy started. Now, more than 120 molecular species have been detected and identified in space. In molecular clouds, besides H_2, these include OH, H_2O, NH_3, CO and many more complex organic (carbon based) ones including formaldehyde, ethyl alcohol, methylamine and formic acid. Recent observations seem to indicate that the amino acid glycine may even be present in these clouds. At least, the molecular precursors (e.g. HCN and H_2O) are known to inhabit these clouds. There are obviously many more molecules to be discovered but detection is more difficult for molecules of greater complexity.

The ISM also includes vast numbers of microscopic solid particles known collectively as interstellar *dust*. They consist mainly of the elements carbon (C) or silicon (Si) with H, O, Mg and Fe in the form of ices, silicates, graphite, metals and organic compounds. The Milky Way contains vast lanes of dust which, being dark, were originally thought to be due to the absence of stars. In fact, dust forms about 1% by mass of all interstellar matter. Most of the dust mass is contained in the larger grains of size exceeding 1000Å, and contain 10^9 atoms. Others are more like large molecules, such as the 'polycyclic aromatic hydrocarbons' (PAHs), consisting of perhaps 100 atoms. By number, most of the grains are actually small (50 Å).

Dust is produced when heavy elements condense out of the gaseous phase at temperatures less than 2000 K. To each element belongs a condensation temperature T_c, at which 50% of the atoms condense into the the solid phase when in thermodynamic equilibrium. For the refractory (rocky) elements (e.g. Mg, Si, Fe, Al, Ca) $T_c \sim 1200$–1600 K; for the volatile elements (O, N, H and C), all critical to life, $T_c < 200$ K.

The origin of the dust is the cool expanding outer layers of evolved red giant stars. These winds are conducive to the condensation of grains from the refractories. The grains are ejected along with gases to contaminate ISM material at a rate of $10^{-7} M_\odot \, yr^{-1}$ per star. The dust is subsequently widely distributed throughout the interstellar medium by the blast waves from supernovae. The dust will cycle several times through diffuse and dense clouds and so becomes well mixed and and processed. In the cold dense clouds, the condensation of other (even volatile) molecules (water, methane etc.) takes place. In molecular clouds, the grains act as nucleation sites for the condensation of even volatile molecules (e.g. water, methane) and the growth of the mantles.

We have discovered numerous inorganic and complex organic molecules in the dense molecular clouds. These molecules may survive in comets and asteroids which could have bombarded the youthful Earth (and other planets) to provide an injection of organic molecules and volatiles (e.g. water) needed for their formation. Some speculation exists that life may even evolve in these clouds and that the Earth was 'seeded' by such a cloud (panspermia).

2.2.2 *Cosmic rays, ions and magnetic field*

Extremely energetic nuclei known as *cosmic rays* penetrate everywhere, processing the ISM through collisions. They mainly consist of protons and electrons with energies that can exceed 10^{20}eV. On collision with hydrogen nuclei, they produce particles called π-mesons which decay into gamma rays. Thus, by measuring the gamma-ray flux, we can constrain the density of hydrogen nuclei. Since we are usually forced to measure cloud mass through trace ingredients, such as CO, cosmic rays provide an important means of corroboration of the H_2 density.

Cosmic rays also penetrate deep into clouds where ionising UV radiation is excluded. In practice, this is at depths which exceed 4 magnitudes of visual extinction (as defined in §2.4.1). The result is that cosmic rays provide a minimum degree of ionisation even in very cold optically thick clouds. This is crucial since the ions interact strongly with both the magnetic field *and* the neutral molecules, binding the field to the fluid. As a result, stars might not form due to the resisting pressure of the magnetic field unless the ions and magnetic field can drift together through the molecular fluid (see §7.5). The ion density n_i is fixed by balancing the ionisation rate in unit volume,

$$\frac{dn_i}{dt} = \zeta \times n(H_2) \quad \text{cm}^{-3}\text{s}^{-1}, \tag{2.1}$$

simply proportional to the number density, with the recombination rate, $5 \times 10^{-7} n_i^2 \, \text{cm}^{-3}\,\text{s}^{-1}$. Here, the coefficient $\zeta = 3 \times 10^{-17}\text{s}^{-1}$ is taken. Equating formation and destruction yields an equilibrium ion fraction

$$\frac{n_i}{n(H_2)} \sim 2.4 \times 10^{-7} \left(\frac{n(H_2)}{10^3 \ \text{cm}^{-3}} \right)^{-1/2}. \tag{2.2}$$

Therefore, an extremely low fractional ionisation is predicted and, indeed, found. Incidentally, the ionisation process also leads to the production of H atoms, which we quantify below in §2.4.3.

An all-pervading but invisible magnetic field exerts a force on electrically-charged particles and so, indirectly, influences clouds of gas and dust. Another effect of the field is to align elongated dust grains. The spin axes of the grains tend towards the magnetic field direction while the long axes orient perpendicular to the field. This effect allows the field direction projected onto the plane of the sky to be deduced. There are two methods. The first method is to measure the polarisation of background starlight. The second method studies the linear polarisation of thermal dust emission. The field direction is transverse to the direction of polarisation, although there are other flow and field effects which may confuse the results. This method often yields a combination of a uniform field and a chaotic structure. Hourglass shapes and toroidal fields have also been reported. The second method is to measure the polarisation of background starlight.

The field strength has proven extraordinarily difficult to measure, mainly relying upon the quantum effect of Zeeman splitting. Molecular lines which have been successfully utilised are CN at 0.3 cm, H_2O at 1.3 cm and OH at 2 cm and 18 cm. The strength and role of the field will be explored in §6.6.

Indirect estimates of the field strength involve large modelling assumptions and, to date, have been applied only to a few, perhaps not typical, cloud regions. These methods involve modelling locations where molecules are excited and compressed in shock waves (as C-shocks – see §10.5), or relate the field strength to the resistance of the field to being twisted or bent in a turbulent medium.

Finally, the ISM is awash with numerous photons of electromagnetic radiation originating from nearby stars as well as a general background from the galaxy as a whole. This radiation is not so pervasive as the cosmic rays or magnetic field but dominates the physics of the cloud edges and provides the illumination for the silhouettes of the dark clouds such as shown in Fig. 2.2.

The cosmic microwave background, the heat of the cooling, expanding Universe, has a temperature $T_{bg} = 2.73$ K. Molecular clouds may indeed get this cold. In the early Universe, at high redshifts (see §13.1.1), the temperature was much higher and microwave background photons inhibited H_2 formation and so probably delayed primordial star formation (see §13.1.1).

Fig. 2.2 EGGs are evaporating gaseous globules emerging from pillars of molecular hydrogen gas and dust. This image of the Eagle Nebula in the constellation Serpens demonstrates the effects of ultraviolet light on the surfaces of molecular clouds, evaporating gas and scattering off the cloud to produce the bright reflection nebula. Extinction by the dust in the remaining dark pillars, resistant to the radiation, blot out all light. Stellar nurseries of dense EGGs are exposed near the tips of the pillars. The Eagle Nebula, associated with the open star cluster M16 from the Messier catalogue, lies about 2 kiloparsecs away. (Credit: J. Hester & P. Scowen (Arizona State U.), HST & NASA).

2.3 Observations

2.3.1 *Radio*

Gas can be excited between specific quantum energy levels and so emit or absorb light at discrete frequencies. Line emission at radio wavelengths can arise from both atoms and molecules. The 21 cm (1420 MHz) line of neutral atomic hydrogen has been used to map out the large scale distribution of atomic clouds. The emission is produced from the ground state of neutral hydrogen where the electron spin axis can be aligned parallel or anti-parallel to the proton's spin axis. When the electron is parallel it is in a higher energy state than when anti-parallel. Mild inter-atomic impacts knock an atom into the higher energy state so that it may then emit upon returning.

Atomic lines are also generated in the radio following a cascade down energy levels after an ion recombines with an electron. Whereas the optical $H\alpha$ line arises from an energetic transition deep within the potential well of the hydrogen atom, other recombination lines originate from very high levels of H as well as other atoms, producing lower energy photons.

Maser emission may be generated when the populations of two energy levels of a molecule become inverted when in a steady state. That is, when the higher energy level also contains the higher population. In this case, a photon with energy corresponding to the transition energy is likely to stimulate further photons moving in the same direction, rather than be absorbed. It follows that the emission can grow exponentially rather than being damped exponentially by absorption. Amplification must be limited by saturation, i.e. it is limited by the rate at which the molecule is pumped into the higher level. Maser spots are so produced, representing those locations where the narrow cone of emission points towards us. H_2O and OH masers are often observed in star formation regions. Other inversions occur in methanol (CH_3OH), ammonia (NH_3) and formaldehyde (H_2CO) and also prove useful tracers of conditions and structure.

Finally, continuum emission is produced as a result of the deceleration of electrons during collisions with ions. This is called free-free emission and is particularly observable at radio wavelengths although the spectrum is much broader.

2.3.2   *Millimetre, submillimetre and far-infrared*

The main constituents, H_2 and He, cannot radiate at the low temperatures of molecular clouds. We, therefore, rely on the ability as collision partners to excite heavier trace molecules such as CO, NH_3 and HCN. These molecules are detected through emission in spectral lines due to transitions in rotational, vibrational and electronic energy states. Downward transitions tend to give radiation in the infrared (vibrational) and submillimetre and millimetre (rotational), whereas electronic changes emit photons in the ultraviolet (UV) and visible range.

The CO molecule is the most abundant tracer with an abundance 10^{-4} times that of the H_2. The CO lines can be employed to determine the temperature. For example, the first three transitions on the rotational (J) ladder emit at 115 GHz (J=1–0), 230 GHz (J = 2–1) and 345 GHz (J = 3–2) and are ideal for tracing cold gas with temperature in the range 5–40 K. The molecule is a linear rotor and is detectable at millimetre wavelengths not only in ^{12}CO but in the ^{13}CO and $C^{18}O$ isotopes also. The energy levels for such diatomic molecules are approximately given by a rigid rotator with $E = B(J+1)J$, where B is a constant, with a selection rule $\Delta J = \pm 1$. This means that the higher-J transitions generate lines in the infrared and highlight warm gas, e.g. the ^{12}CO J = 19–18 transition at 140 μm will produce most CO emission for gas at 1000 K while the J = 6–5 transition at 434 μm represents the peak for 100 K gas. In this manner, temperatures of clouds can be probed through the ratios of low-J and mid-J rotational CO intensities as well as numerous other molecular line ratios.

The density can be constrained by measuring emission from different molecules. This is because some molecules are more prone to collisional transitions and others to radiative transitions. We demonstrate this by taking a highly simplistic model. The average distance that a molecule will travel before colliding with another molecule is called the mean free path, λ_p. This distance will be shorter in denser regions or where the molecules present larger cross-sectional areas. Therefore, we expect

$$\lambda_p = \frac{1}{\sigma_p \, n(H_2)} \tag{2.3}$$

where σ_p, the collision cross-section, is related to the impact parameter (approximately the molecule size) of order of 10^{-15} cm^2. The rate of collisions of a molecule with the H_2 is then the inverse of the collision timescale t_C

$$C = \frac{1}{t_C} \sim \sigma_p n(H_2) v_{th} \tag{2.4}$$

where the mean velocity can be approximated as $v_{th} \sim 10^4 \sqrt{T}$ cm s^{-1}. Hence the collision or excitation rate is roughly $C = 10^{-11} n(H2) \sqrt{T}$ s^{-1} for all molecules. We now compare this to the known de-excitation rate through spontaneous emission, A, which is the probability per unit time for radiative decay. If $C << A$ then excitations are relatively inefficient. The cloud density at which the molecule is likely to radiate most efficiently and, hence, the critical density at which it serves as a tracer is where $C \sim A$. Taking a cold cloud of temperature 10 K then gives the following.

The CO 1–0 line at $115\,\mathrm{GHz}$ (with $A = 7 \times 10^{-8}\,\mathrm{s^{-1}}$) traces low density molecular clouds with $n = 10^2 - 10^3\,\mathrm{cm^{-3}}$ and is, therefore, the choice for mapping galactic-scale molecular distributions. On the other hand, CS 2–1 at $98\,\mathrm{GHz}$ (with $A = 2 \times 10^{-5}$) traces high density molecular clumps with $n \sim 10^6\,\mathrm{cm^{-3}}$.

Although CO is easily excited, radiative transfer will affect the detectability. If the opacity is high, the emitted photon will be reabsorbed, reducing the value of A. Self-absorption due to lower excitation foreground gas will also dramatically distort observed line profiles. To measure the optical depth, we utilise lines from different isotopes. For example, if the 1–0 lines from both ^{12}CO and ^{13}CO are optically thin, then the intensity ratio is given by the abundance ratio of *sim* 90. If we find the measured ratio is considerably lower, then we can conclude that the ^{12}CO line is optically thick.

Certain atoms and ions possess fine structure in their electronic ground state. This results from coupling between orbital and spin angular momenta. The splitting is typically 0.01–$0.1\,\mathrm{eV}$ and the upper levels are relatively easily reached through collisions in quite warm dense gas. Hence, provided the atoms are not bound onto dust or molecules, they can trace physical conditions in clouds. In particular, the C I line at $609\,\mu\mathrm{m}$, the singly-ionised C II line at $158\,\mu\mathrm{m}$ and the O I line at $63\,\mu\mathrm{m}$ are often prominent in far-infrared spectra.

2.3.3   Infrared observations

Molecular hydrogen can be directly observed in the infrared. Unfortunately, it can only be observed when in a warm state or exposed to a strong UV radiation field. Being a symmetric molecule, dipole radiation is forbidden and only quadrupole transitions with rotational jumps $\Delta J = 2$ (or 0, of course, for a vibrational transition) are allowed. This implies that there are two types or modification of H_2: ortho (odd J) and para (even J). The difference lies in the nuclear spins which are only changed in certain types of collision. For ortho H_2, the nuclear spins are aligned to yield a spin quantum number of $S = 1$ while for para H_2 the spins are anti-parallel and $S = 0$. Quantum mechanics then yields the number of different discrete states that the molecule can exist in as proportional to $(2J+1)(2S+1)$. For this reason, ortho lines are often found to be a few times stronger than para lines when a sufficient range of levels are occupied. On the other hand, the least energetic transition is in para-H_2 between $J = 0$ and $J = 2$, with an energy of 510 K. For this reason, para H_2 will be dominant in cold gas in which the molecule approaches equilibrium.

The most commonly observed H_2 lines lie in the near-infrared. Water vapour in the atmosphere contributes substantially to the opacity leaving specific wavelength windows open to ground-based telescopes. These are centred at wavelengths of $1.25\,\mu m$ (J-band), $1.65\,\mu m$ (H-band) and $2.2\,\mu m$ (K-band). These transitions are ro-vibrational which means that changes in both vibrational and rotational levels are permitted. The most commonly observed spectral line is the 1–0 S(1) transition at $2.12\,mum$, used to image hot molecular gas where protostellar outflows impact with their molecular cloud. The 1–0 denotes the vibrational change from the first to the ground level, while the S(1) denotes the change in rotational state with S denoting $\Delta J = +2$ and the final state is $J = 1$ (the letters Q and O traditionally denote the other permitted transitions $\Delta J = 0$ and -2, respectively).

Warm dust particles emit light across a broad continuous range of infrared frequencies, producing a 'continuum' given by Kirchoff's laws. The spectrum will peak at frequencies proportional to the temperature of the dust, as quantified in §6.1. Most bright adult stars tend to emit in the visible and above because of their higher temperature. Hence dust generally stands out in the infrared bands.

2.3.4 *Optical, ultraviolet and X-rays*

Dense atomic gas which is ionised either by radiation or by collisions in shock waves will subsequently recombine. The recombination spectrum consists of numerous well-studied lines, generated as the recombining electron cascades down the energy ladder. In particular, the $H\alpha$ Balmer line of H (from the $n = 3$ to the $n = 2$ orbit, where n is the principle quantum number) at $6563\,\text{Å}$ is mainly responsible for the red colour of many so-called reflection nebula. The Lyman-alpha line $Ly\alpha$, at $1216\,\text{Å}$ (from $n = 2$ to the ground $n = 1$), is significant as a coolant on top of contributing significantly to the ultraviolet.

Cold interstellar gas can be detected through atomic hydrogen absorption lines (in the visible), and to a lesser extent the molecular hydrogen absorption lines (in the UV which is efficiently absorbed). Of course, we require the nebula to be back-lit by a star and for some of the stellar light to survive the passage through the cloud. These lines can be distinguished from the star's lines by the fact that they will be quite sharp since the gas is much cooler.

X-ray emission is not observed from molecular clouds. Hot gas in the interstellar medium and in the vicinity of young stars emits Bremsstrahlung emission, which is a form of free-free emission as described in §2.3.1.

2.4 Processes

2.4.1 *Interstellar extinction and reddening*

Dust is recognised as the substance which makes Dark Clouds, such as the Horsehead Nebula, dark. The dust obscures light from objects in the background as well as from internal objects. Dust can be detected due to four primary effects that it has on starlight. These are Extinction, Reddening, Polarisation and Infrared Emission.

Dust extinction is the dimming of starlight caused by absorption and scattering as it travels through the dust. The ability for photons to penetrate is proportional to the column of dust in a cloud, which is proportional to the column density of gas, the number of hydrogen atoms per unit area, in its path, for a given ratio of dust to gas particles. The extinction grows exponentially with the column: we find that a layer of gas with a column of $N_H = 2 \times 10^{21}$ cm^{-2} usually contains enough dust to decrease the number of visual photons by a factor of 2.5. We term this factor a *magnitude* of extinction. This is evident from decades of observations of the extinction cross-section, summarised in Fig. 2.3. A further column of gas thus reduces the remaining radiation by another factor of 2.5. Therefore, a column of gas of 10^{23} cm^{-2} reduces the number of photons by fifty 'magnitudes' of 2.5, or a factor of 10^{20}. Such columns are often encountered in star-forming cores, obliterating all visual and UV radiation from entering or escaping. To summarise, the visual extinction typically indicates the column of interstellar gas according to

$$A_V = \frac{N_H}{2 \times 10^{21} \text{ cm}^{-2}}. \tag{2.5}$$

Fortunately for us, extinction is not equal at all wavelengths – it is now often possible to see through the dust thanks to advances in radio and infrared astronomy. This is illustrated in Fig. 2.3. For example, the near-infrared extinction at 2.2μm in the K-band is related to the visual extinction by $A_K \sim 0.11\,A_V$. An infrared extinction law of the form $A_\lambda \propto \lambda^{-1.85}$ is often assumed. Dust emission and absorption bands generate prominent features in the infrared spectra. A well-known bump at $9.7\,\mu$m is attributed

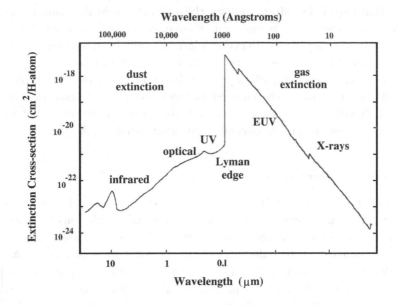

Fig. 2.3 A schematic diagram displaying interstellar extinction across the spectrum. Note the wide range in extinctions and the advantage of infrared observations over optical and ultraviolet (from original data accumulated by Ch. Ryter).

to a silicate feature from Si–O stretching. A $3.07\mu m$ feature is accredited to water ice. Other ice features include CO and NH_3. It appears that the volatiles condense out in dense regions onto the resilient silicate and graphite cores. Other previously 'unidentified bands' appear to be caused by Polycyclic Aromatic Hydrocarbons (PAHs), discussed in §2.2.1.

Reddening occurs because blue light is more strongly scattered and absorbed than red. This is quantified by the colour excess produced between any two wavelengths. Measured in magnitudes, the excess is proportional to the extinction provided the dust properties in the ISM are uniform. For example, we can derive the selective extinction $R = A_V/E_{B-V} = 3.1$ as characteristic of the diffuse ISM where E_{B-V} is the relative excess produced by extinction between the optical filters B $(0.44\mu m)$ and V $(0.55\mu m)$. In dark clouds such as ρ Ophiuchus, $R \sim 4.2$, indicating the presence of larger grains, due to coagulation or to the accretion of volatiles onto grain mantles.

Starlight can become polarised when passing through a dust cloud. Elongated grains tend to have their spin axes aligned with the magnetic field, producing polarisation along the magnetic field direction. The reduced extinction in the infrared allows the magnetic field direction to be traced in dense clouds. Scattering even off spherical grains can also polarise light. In this case the polarisation is orthogonal to the local magnetic field.

Finally, the dust particles are responsible for much of the observed infrared radiation. Each grain will absorb visible and UV light from nearby stars, heat up and emit in the infrared as much energy as it absorbs. The IRAS and COBE satellites showed that infrared emission was strongest in regions where there is a high concentration of interstellar gas.

2.4.2   *Photo-dissociation*

For molecular clouds to survive, the molecules need to be protected from dissociation by high energy photons. Extreme UV (EUV) photons from hot O and B stars are unlikely to reach the molecules because they ionise atomic hydrogen. The EUV is also called the Lyman continuum, with energies $h\nu > 13.6\,\text{eV}$, equivalent to wavelengths $\lambda < 912\,\text{Å}$, sufficient to photo-ionise even cold atomic hydrogen from its ground state. The wavelength $\lambda = 912\,\text{Å}$ is defined as the Lyman limit or 'edge'. The dramatic effect of this edge is evident in Fig. 2.3. The extreme UV creates H II regions around massive stars which completely absorb the EUV and re-radiate it at longer wavelengths. This shields the molecules from the EUV.

Molecules directly exposed to far-UV radiation, however, are also rapidly destroyed. Far UV radiation with energy in the range 5–$13.6\,\text{eV}$ (912–$2000\,\text{Å}$) is able to penetrate the skin of a cloud. The photons ionise atoms such as carbon and iron and photo-dissociate hydrogen molecules (but do not ionise). The transition layer between the exposed dissociated gas and the molecular interior is not smooth but probably very wrinkled and clumpy, allowing radiation to penetrate quite deep. We call this thick skin a *Photo-dissociation Region* or PDR.

A molecule is dissociated in two steps. It first undergoes electronic excitation by absorption of a resonant UV photon. Then, there is about a 10% probability that the molecule radiates into a state in the vibrational continuum, in which the two atoms are not bound and so fly apart. Alternatively, and most probably, dissociation does not result but the excited molecule returns to the ground electronic state where it cascades down through the vibrational and rotational energy states. This generates a 'fluorescence spectrum' of H_2 emission lines, characterised by many strong lines originating from high levels of vibration.

For molecules to build up they require their absorption lines to become optically thick, thus forming a self-protecting layer to the far UV. This is called self-shielding. In the absence of self-shielding, the radiation will still be attenuated by dust, and molecules will form when sufficient dust lies in the path of the radiation.

For H_2 self-shielding to be effective, however, we require only the equivalent of about $A_V \sim 0.1$. The CO molecule, being less abundant and so less effective at self-shielding, requires $A_V \sim 0.7$. Thus a dense cloud exposed to UV radiation will possess several layers of skin. The full list of layers is as follows:

- an outer ionised hydrogen region,
- a thin neutral atomic H layer,
- an outer H_2 layer containing C and O atoms,
- an inner H_2 layer containing CO, and
- a dark core containing H_2, CO and O_2.

Most of the PDR radiation originates from the warmed C and O atoms through their fine-structure emission lines (see §2.3.3).

2.4.3 *Hydrogen chemistry*

Molecules do not form readily in interstellar gas. Collisions between atoms are too fast for them to lose energy and become bound: a dynamical interaction time is typically just 10^{-13} s whereas the time scale to lose energy via radiation is at least 10^{-8} s. Furthermore, three-body collisions in which the third body could carry away the excess energy are extremely rare. Besides these problems, the mean UV photon flux in the Galaxy is of order 10^7 photons $cm^{-2}\,s^{-1}$ and a molecule presents a cross-section of order $10^{-17}\,cm^2$ to the passing photons. Hence, exposed molecules survive for periods of just 10^{10}s, just 300 yr.

Molecular hydrogen forms much more efficiently on the surfaces of grains. The main requirement is that one atom is retained on the grain surface until a second atom arrives and locates it. At low gas and grain temperatures and with a typical distribution of grain sizes, we usually take a rate of formation per unit volume of $3 \times 10^{-18}\,n\,n(H)\,T^{1/2}\,cm^{-3}\,s^{-1}$. Hence, the time scale for H_2 formation, normalised to a typical molecular clump, is

$$t_F = 3 \times 10^6 \left(\frac{n}{10^3\ cm^{-3}}\right)^{-1} \left(\frac{T}{10\ K}\right)^{-1/2} \text{ yr.} \qquad (2.6)$$

We will later find that this is also a typical cloud age.

High uncertainties, however, concern the nature of the dust and the limitations of laboratory experiments. In particular, the dust temperature, T_d, is a critical factor. If T_d is less than about 20 K, then H_2 formation might proceed at the above rapid rate. Above 20 K, the formation rate appears to decrease enormously. In comparison, observational estimates of dust temperatures in the diffuse unshielded ISM range from 13 K to 22 K. In star formation regions, however, the stellar radiation field provides considerable dust heating. An estimate of how rapidly the dust temperature is reduced with depth through a cloud surface is given in terms of the visual extinction A_V from the formula $T_d^5 \propto \exp(-1.66\,A_V)$.

The above H_2 formation time is still clearly insufficient to form molecular clouds until the cloud is shielded from the UV radiation. When sufficient, however, the cloud will become fully molecular with cosmic rays maintaining a tiny fraction of H atoms: balancing molecule destruction at the rate $\xi\,n(H_2)$ with the above formation rate yields a density of hydrogen atoms of $0.1\,cm^{-3}$ within dark clouds, independent of the cloud density. Molecules will also be dissociated by UV radiation produced internally by collisions with the cosmic ray-induced electrons.

Molecular ions play a key part in cloud chemistry since they react fast even at low temperatures. H_3^+ is a stable ion produced by the cosmic rays. Along with CO^+, it is often important in the subsequent ion-molecule chemistry.

What if no dust is present, such as in primordial gas (§13.1.2) and probably also in protostellar winds and jets (§10.7)? Despite the above arguments, molecular hydrogen will still form in the gas phase under certain conditions. It is a two-step process. We first require H^- to be produced via the reaction $H + e \rightarrow H^- + h\nu$. Then, some of the H^- can follow the path $H^- + H \rightarrow H_2 + e$, although the H^- may be, in the meantime, neutralised by reactions with protons or the radiation field. A complete analysis shows that a rather high electron fraction or a paucity of dust is necessary for these reactions to assume importance. It should also be noted that in collimated protostellar winds, the density may be sufficient for three-body formation: $3H \rightarrow H_2 + H$.

Besides photo-dissociation discussed above, H_2 is destroyed in collisions with H_2 or other molecules and atoms. Each collisional dissociation removes 4.48 eV, the molecular binding energy. Hence dissociation also cools the gas while reformation may then heat the gas (although the reformed excited molecule may prefer to radiate away much of this energy before it can be redistributed).

Under what condition will hydrogen return to the atomic state simply through energetic collisions? During the collapse of a protostellar core, the gas warms up and molecule dissociation becomes the most important factor (§6.6). Approximately, the dissociation rate (probability of dissociation per second) takes the form $5 \times 10^{-9} \, n(H_2) \exp(-52,000 \, K/T) \, s^{-1}$ (note that 4.48 eV has converted to 52,000 K). This equation holds for high density gas since it assumes that all energy levels are occupied according to collisions (thermal equilibrium) and not determined by radiation. Using this formula, if gas at a density of $10^6 \, cm^{-3}$ is heated to 1,400 K, it would still take $3 \times 10^{18} \, s$, to dissociate. That is much longer than the age of the Universe. At 2,000 K, however, the time required is down to just $3 \times 10^{13} \, s$, or a million years – comparable to the collapse time. Therefore, if energy released during a collapse is trapped in the gas, a critical point may be reached where molecule destruction is triggered. This is exactly what we believe occurs (see §6.6).

2.4.4 *Chemistry*

Astrochemistry provides a means to connect the structure and evolution of molecular clouds to the distributions and abundances of numerous molecular species. It usually requires the computational treatment of a complex network of chemical paths. Chemical models have now been developed to predict the existence and abundance of various molecules. The predicted abundances of some main species are in agreement with observed values but for many other molecules, especially heavy atom-bearing molecules and large polyatomic molecules, anomalies persist. Revisions to the models are ongoing, including improved treatments of the neutral-neutral reactions and grain-surface molecular depletion and desorption. As a result, the relative column densities of observed molecules can be employed to infer the chemical evolution and may serve as an indicator of cloud history.

We summarise here the facts most relevant to an overview of star formation. The chemistry can be broken up into three categories.

Shielded cloud chemistry applies if the gas is cold and photodissociation is low. This should be relevant to the dense inner regions of dark clouds. Cosmic rays still penetrate and they provide the catalyst for the chemical processing. Almost all the hydrogen is in molecular form so the reaction network begins with the cosmic ray production of H_2^+. At the rate of ionisation of $10^{-17} \, \zeta \, s^{-1}$ from §2.2.2, the average H_2 is ionised roughly once every 10^9 yr. Once ionised, the H_2^+ quickly reacts with another H_2 to form H_3^+ and an H atom. This ion-neutral reaction proceeds swiftly, according to a so-called Langevin rate constant of 2×10^{-9} cm^3 s^{-1}. This implies that in a medium of density 1000 cm^{-3}, the average wait is several days to react to form H_3^+. Clearly, the initial ionisation is the rate-limiting step.

The chemistry now all turns on the H_3^+. The many processes include proton transfer with carbon and oxygen atoms (e.g. $C + H_3^+ \rightarrow CH^+ + H_2$), dissociative recombination (e.g. $HCO^+ + e \rightarrow CO^+ + H_2$, generating CO) and radiative association (e.g. $C^+ + H_2 \rightarrow CH_2^+ + \nu$, where e represents an electron and ν a photon. Heavy elements are also removed from the gas phase by condensation onto dust grains.

Exposed cloud chemistry applies to molecular regions of low extinction. Here, CO is photo-dissociated and the carbon is photo-ionised by starlight since it has an ionisation potential of 11.26 eV, below the Lyman limit. Therefore, C^+ is the most abundant ion with a fraction $n(C^+) \sim 10^{-4} n(H_2)$, which far exceeds the dense cloud level given by Eq. 2.2. The C^+ is converted to CO^+, which then is converted to HCO^+, which produces CO through dissociative recombination as noted above. The CO will take up all the available carbon when self-shielded to the ultraviolet radiation. Otherwise, the CO is rapidly dissociated through $CO + h\nu \rightarrow C + O$.

Warm molecular chemistry occurs when the gas is excited by collisions to temperatures of order 1,000 K. The sequence $O + H_2 \rightarrow OH + H$ - 4480 K and then $C + OH \rightarrow CO + H$ is very rapid in hot gas, usually converting C to CO before cooling reduces the temperature which would slow down the reactions. Note that the process is endothermic: collisional energy is required to make the OH.

Water molecules also follow from the OH through another endothermic reaction $OH + H_2 \rightarrow H_2O + H$ - 2100 K. In cold dense cores or collapsing cores, many molecules including H_2O accrete onto grains, forming icy mantles. Surface chemistry and radiation processes modify the composition.

The carbon monoxide molecule is the best tracer of star-forming gas on large scales for several reasons. First, it is rotationally excited even in gas of a few Kelvin. Second, it is formed very efficiently. Thirdly, it is not easily destroyed with a relatively high dissociation energy of 11.09 eV (electron Volts). Nevertheless, CO is also often depleted in dense regions, forming CO ice. It is quite commonly depleted by an order of magnitude or more, along with other species such as CS and HCO^+.

In the presence of a young interacting star, the ices are heated and molecules evaporated from the dust. This occurs in a sequence, according to their sublimation temperatures. Shock waves and turbulence will also strongly heat the cloud on larger scales, returning volatiles and radicals back into the gas phase through processes such as high-energy ion impacts, termed sputtering. Shattering also occurs through grain collisions. This tends to modify the distribution of grain masses.

It should also be mentioned that large enhancements of minor isotopes have been recorded in dense regions near young stars. Examples are DCN and DCO^+. These are explained as due to deuterium fractionation in which the deuterium initially in HD molecules undergoes a path of reactions.

2.4.5   *Cooling and heating*

The capacity of molecular clouds to cool is absolutely crucial to the mass of stars which form (see §4.3.4 on the Jeans Mass). Radiation is thought to be the chief carrier of the energy from a cloud (rather than conduction, for example). Photons are emitted from atoms and molecules as the result of spontaneous de-excitation. The particles are first excited by collisions, mainly through impacts with H_2, H, dust or electrons. If the density is high, however, the de-excitation will also be through an impact rather than through a photon, as explicitly shown in §2.3.2). This implies that radiative cooling is efficient at low densities but impacts redistribute the energy at high densities. We define a critical density to separate these two regimes (referred to in the literature as local thermodynamic equilibrium, or LTE, above, and non-LTE below). This concept is probably vital to primordial star formation (§13.1.3).

An important coolant is CO through rotational excitation as well as trace atoms C II at low densities and neutral C and O fine-structure excitation at high density. Although dominant by number, H_2 is not usually efficient at radiating away energy since it is a symmetric molecule in which a quadrupole moment is not effective.

In warm gas, the molecules CO, OH and H_2O also make considerable contributions to the cooling, and vibrational H_2 emission cools molecular gas after it has been strongly heated in shock waves to over $1,000\,K$. Rotational excitation of H_2 will also cool gas heated above $200\,K$.

A second important cooling path is available through gas-grain collisions. Provided the grains are cooler than the gas, collisions transfer energy to the dust. The dust grains are efficient radiators in the long wavelength continuum (in the infrared and submillimetre) and so the radiated energy escapes the cloud. Clearly, the dust is also a potential heating source through collisions if the dust can be kept warmer than the molecules by background radiation.

Cloud heating is provided by far ultraviolet photons. They eject electrons at high speed from grains. The excess energy is then thermalised in the gas. This process is appropriately called *photoelectric heating*. Deep inside clouds where the FUV cannot penetrate, heating is provided through the cosmic ray ionisations discussed above with perhaps 3.4 eV of heat deposited per ionisation.

Gas motions also heat the gas. The energy in subsonic turbulent motions is eventually channelled by viscosity into thermal energy after being broken down into small scale vortices (as discussed in §4.4). The energy of supersonic turbulence is dissipated more directly into heat by the creation of shock waves, described in §4.4. In addition, the gravitational energy released by cloud contraction produces compressive heating (or cooling in expanding regions).

There are many other potential heating mechanisms that we have already come across, depending on the state of the gas. These include direct photo-ionisation, collisional de-excitation of H_2 after UV pumping and H_2 formation.

2.5   Summary

This chapter has served three purposes: to provide backgrounds to the materials, to the means of observation and to the physical processes. Particular regard has been given to the processes relevant to molecular clouds. In fact, it is the large number and variety of physical processes which creates a fascinating story. While the cast of characters is small, they can take on many roles. Each molecule and atom has its own peculiarities and regime of importance, either physically, observationally or both.

Having explored the means by which light is emitted, absorbed and scattered, we are ready to appreciate and interpret the observations. The following chapters will then focus on the dominant interactions which describe the conditions suitable for star formation.

Chapter 3

The Clouds

There is overwhelming evidence that stars are born inside clouds. Yet it was not always obvious that this had to be so. Stars could have been eternal beacons in a steady-state Universe. They could have come into existence in the early Universe, or simply built up through collisions and coagulation of clouds of atoms.

Although normal stars like our Sun consist of atoms and the gas from which they originate was also atomic, stars are born in *molecular* clouds. More precisely, *all present day* star formation takes place in molecular clouds. These protected environments also serve as the wombs and the nurseries for the young stars. It has thus become clear that the processes in molecular clouds hold the key to understanding star formation. Without this intermediate molecular stage, galaxies would be very different.

Most of a cloud consists of hydrogen molecules, H_2. To observe it, however, we require a surrogate since the H_2 is simply too cool to be excited and so virtually inaccessible to direct observation. We now have the technology to detect the emission of many trace molecules and to penetrate deep into the clouds. The detection of interstellar ammonia in 1968 indicated the existence of very dense clouds. We can now even map the distributions of various molecules and distinguish collapsing molecular cores.

What we have discovered has drastically altered our view of star formation. We shall show that clouds are much younger than previously thought. Secondly, many clouds and clumps are a figment of observations with limited dynamic range combined with the human tendency to split complex patterns into recognisable units. The apparent structures often represent a convenient means of categorisation rather than real entities. First, we need to assimilate the fundamental evidence relating to cloud origin, age and internal structure.

3.1   Phases of the Interstellar Medium

How much material do we have available and in what form? Molecules
and molecular clouds form out of an expansive reservoir which occupies
over 99% of the volume. This is the atomic interstellar medium. The
gas is atomic because the ultraviolet light from massive stars dissociates
any molecules much faster than they can reform (as discussed in §2.4.3)
Only where material is protected from the UV can molecular clouds, and
ultimately stars, form.

We recognise four atomic components, two of which are largely ionised
gas and two neutral, as listed in Table 3.1. First, there is a hot ionised
component. Although hot and buoyant, its low density and pressure stops
it from blasting away the entire interstellar medium. Related hotter gas,
termed coronal gas, with temperatures of millions of Kelvin and particle
densities under $0.01\,\mathrm{cm}^{-3}$ is observed in soft X-rays. It is heated by blast
waves from supernovae and violent stellar winds. This gas escapes, often
diverted into chimneys which funnel it out of the galactic plane.

The second component is a warm ionised medium of lower temperature
and higher density, but still very diffuse. The electron fraction is close to
unity. The ionised components together could occupy up to about 80% of
the interstellar volume, although these numbers are very uncertain.

One neutral component consists of warm gas with temperature $\sim 8000\,\mathrm{K}$
and density $\sim 0.3\,\mathrm{cm}^{-3}$. It is composed largely of neutral hydrogen (i.e. H I,
so observable in the 21 cm line) with about 2–20% of ionized gas, including
electrons (so observable as free-free in the radio continuum, see §2.3.1). It
probably fills about 20% of the volume in the disk of the Galaxy. The
stability of this phase is maintained by photoelectric heating and by Lyα
and recombination cooling, which together act as a thermostatic regulator.

Finally, there is a cold atomic medium with temperature under $100\,\mathrm{K}$,
density 30–$50\,\mathrm{cm}^{-3}$ and a very low ion fraction (under 0.1%). Photoelectric
heating is balanced here by C II and O I cooling in the fine-structure lines
(see §2.3.2).

The densest parts of the ISM are observed as molecular clouds. These
are just parts of dense condensations within the atomic phases. Apparent
sharp boundaries are exactly that: apparent. The boundaries are not necessarily edges in density. Instead, they are boundaries in the phase, with
extended envelopes of atomic gas around the clouds.

The phase transition takes place where the ultraviolet radiation flux has
reached the limits of its penetration into the accumulated gas. The ability

to penetrate is measured by the column of shielding gas. When sufficient, this is like a thick skin to the cloud, the degree of penetration being given by the column density of the absorbing dust in the skin.

The clouds are classified, according to their opacity, into diffuse, translucent and dark clouds; and by size, into Giant Molecular Clouds (GMCs), clumps and cores (see Table 3.2). Translucent clouds are those in which their opacity (the optical depth from the edge to the centre) is of order of one to a few at visual wavelengths. They have drawn particular attention since the physical processes and interaction with external radiation are spread out over large distances. We can thus conveniently measure variations of quantities with distance into the cloud and so test our models.

The total molecular mass in our Galaxy of $3 \times 10^9 \, M_\odot$ is as large as the atomic mass. It is not distributed evenly but 90% is closer to the centre of the Galaxy than the Sun (inside the 'solar circle'), in comparison to only 30% of the atomic gas. Note that almost all the mass of gas is cold and occupies a small volume.

Table 3.1 Phases of the interstellar medium. These phases may not be in equilibrium but dynamic and short-lived at any particular location.

Phase	Temperature K	Density cm^{-3}	Fraction of Volume	Mass in $10^9 \, M_\odot$
Hot ionised medium	$3\text{--}20 \times 10^5$	3×10^{-3}	0.4–0.7	0.003
Warm ionised medium	10,000	3×10^{-1}	0.15–0.4	0.05
Warm neutral medium	8000	4×10^{-1}	0.2–0.6	0.2
Cold neutral medium	40–100	6×10^1	0.01–0.04	3
Molecular Clouds	3–20	3×10^2	0.01	3

3.2 Weighing up Molecular Clouds

Dense cool gas in molecular form occupies just 2–4% of the interstellar volume. It is mainly contained within giant molecular clouds (GMCs), typically tens of parsecs in size and containing up to a million solar masses. In fact, for some unknown reason, there appears to be a limit to a cloud mass of six million solar masses. The properties listed in Table 3.2 are broad estimates.

Table 3.2 Estimated properties of individual molecular aggregates in the Galaxy.

Phase	GMCs	Clumps/Globules	Cores
Mass ($M_\odot$)	6×10^4–2×10^6	10^2	1–10
Size (parsecs)	20–100	0.2–4	0.1–0.4
Density (cm^{-3})	100–300	10^3–10^4	10^4–10^5
Temperature (K)	15–40	7–15	10
Magnetic Field (μG)	1–10	3–30	10–50
Line width ($km\,s^{-1}$)	6–15	0.5–4	0.2–0.4
Dynamical life (years)[a]	3×10^6	10^6	6×10^5

[a] Note: dynamical life defined as Size/(Line Width), true lifetimes would be considerably longer if clouds were static.

There are many sources of inaccuracy in the parameters of individual clouds. The mass, temperature and internal motions (velocity dispersion) rely on an interpretation of molecular line observations. The size depends on the measured angular size and a distance estimate. However, the cloud distances are only approximate and even the nearby examples may be 20% closer or further away than assumed.

Accurate distances are required to derive accurate sizes, luminosities and masses of both clouds and young stars. We can classify the methods employed to estimate cloud distances but remark that all three classes are indirect: the distance to stars is actually determined.

The classical methods employ the obscuration of the cloud to distinguish background stars, foreground stars and associated stars (e.g. as reflection nebulae). If we knew the distances to these stars, then we could place the cloud. We achieve this with spectroscopic observations of the individual stars to determine their spectral type. Once certain types of stars are classified, we know their absolute luminosity and so can determine their distances. Finally, knowing their distance and whether they are foreground or background is sufficient to restrict the cloud distance.

Astrometric methods are now capable of determining distances to bright young stars associated with the molecular clouds. The Hipparcos satellite used its orbit around the Sun to determine distances by the method of parallax (the apparent motion of nearby objects relative to distant objects). However, milliarcsecond accuracy is not sufficient to obtain a precise distance even to the nearest star formation regions.

The third class of methods employs independent determinations of speed. Given an angular speed and a true speed, we can obtain a distance. The angular speed can be derived from an observed motion of an object or shock wave in the sky (proper motion) or the rotation period of spots on the surface of a young star. The true speed can be derived from modelling the broadening of emission lines in terms of Doppler shifts. These methods require some geometrical or statistical assumption to relate the two speeds and extract the distance.

Molecular clouds are routinely mapped and weighed in the CO rotational transition at 2.6 mm (since the H_2 is not detectable). Empirical relations have been deduced to relate CO and H_2 including one for the H_2 surface density $\Sigma(H_2)$ in terms of the CO J=1–0 intensity, I(CO): $\Sigma(H_2)$ = 5 ± 3 I(CO) $M_\odot \, pc^{-2}$. The derived quantity is the surface or column density i.e. the total H_2 through the cloud along the line of sight. The CO line is usually optically thick and observers often resort to mapping weaker emission from optically-thin isotopes. In all cases, an abundance for the trace isotope must be justified. Cloud masses are then estimated by integrating the column of gas over the cloud extent. Finally, the cloud density is estimated after assuming a reasonable size for the cloud along the line of sight.

Dense cores have been mapped in other molecules such as NH_3 or HCN. Occasionally, we can estimate a gas density with this approach since different molecules and different energy levels within a molecule require particular critical densities for excitation (§2.3.2).

Dust emission can also be used to determine cloud masses although the dust emissivity and temperature need to be specified. In the millimetre or submillimetre continuum the dust emission is typically optically thin and the emission from thermal dust is given in the simple low-frequency Rayleigh-Jeans limit (see §6.1). This method is commonly employed to determine masses of dense cores since tracing molecules become unreliable as they are heavily depleted from the gas phase into the dust itself. In turn, this alters the dust emissivity and this must be taken into account (see §6.1). Thus, while a single mass determination method relies on several assumptions, one can build confidence through a broad corroborating approach.

An alternative technique has been successfully pioneered which employs dust extinction. We can now measure the extinction to up to thousands of stars behind a cloud due to the recent deployment of large and sensitive infrared array detectors. The extinction can then be scaled to a column density. Working in the infrared, we are able to detect sufficient stars in the background even where there are 25 magnitudes of visual extinction. As a result, we can also map out how the dust is distributed.

One can also calculate a cloud mass, M_{vir}, by assuming *virialisation* or *virial equilibrium* instead of assuming an abundance. This requires knowledge of the internal motions. The basic argument is that if the motions in a cloud were not precariously held in equilibrium by gravity then the cloud would either disperse or collapse in a brief time. We begin by observing an emission line and fitting a Gaussian line shape to its profile of the form

$$ I(v) \propto \exp\left[-\frac{v^2}{2\sigma_v^2} \right] \tag{3.1} $$

to determine the velocity dispersion, σ_v. The classical formula for the escape speed v_{esc} from the surface of a spherical cloud, planet or star of mass M and radius R is

$$ v_{esc} = \sqrt{\frac{2\,G\,M}{R}}. \tag{3.2} $$

A virialised cloud, in which the kinetic energy of cloud motions is not sufficient for material to escape yet enough to stabilise against collapse, is the most likely condition for observation. The virial theorem will be discussed in §4.3.1. Here, we do not actually require a static equilibrium but, at least, a temporary energy equipartition. Whether true or not, we can still calculate a theoretical 'virial mass' for any cloud, given an observed velocity dispersion (i.e. from the motions along our line of sight). For a uniform cloud of radius R_c, we find a velocity dispersion

$$ \sigma_v(0) = \sqrt{\frac{G\,M_{vir}}{5\,R_c}}, \tag{3.3} $$

significantly lower than the surface escape speed. Reversing, we derive the virial mass given the velocity dispersion:

$$ M_{vir} = 1160 \left(\frac{R_c}{1 \text{ pc}} \right) \left(\frac{\sigma_v}{1 \text{ km s}} \right)^2 \text{M}_\odot. \tag{3.4} $$

Clouds may develop evolved cores in which the gas density rises steeply towards the centre. Given a spherical cloud with a density structure

$$ \rho = \rho_o \left(\frac{R}{R_c} \right)^\alpha, \tag{3.5} $$

it can be shown that the velocity dispersion increases by a factor

$$\sigma_v(\alpha) = \sqrt{\frac{5}{3}\frac{(3+\alpha)}{(5+2\alpha)})}\, \sigma_v(0),\tag{3.6}$$

quite close to unity even in the least uniform cores with $\alpha = -2$.

In practice, it is straightforward to measure the full line width at half the peak intensity, Δv, and then apply the formula for the Gaussian distribution to obtain $\sigma_v = 0.426\, \Delta v$.

The virial method, however, can be very misleading because clouds are, in general, not bound. The evidence will be given in §3.4.

3.3 Molecular Clouds

3.3.1 *Breeding grounds for high-mass stars*

Molecular clouds are found throughout our Galaxy, including the Galactic centre and at quite high latitudes from the Galactic plane. They are, however, very strongly associated with the Galactic plane and preferentially associated with the spiral arms. The locations of major regions are indicated in Fig. 3.1.

Millimeter wavelength CO maps show the typical structure of a cloud: inhomogeneous, elongated and containing dense clumps and filaments. The dense clumps are very closely related to regions of on-going star formation.

One major debate is whether star formation occurs spontaneously or whether it is triggered. We first demonstrate with some examples that the formation of high-mass stars is initiated by various external influences due to neighbouring active phenomena such as shock waves from supernova remnants, expansion of H II regions, and UV radiation from other massive stars.

3.3.2 *Orion*

Messier 42, also known as the Great Orion Nebula, is perhaps the most well-known and well-studied star-forming region. It extends 15° across the sky and lies conspicuously well below the Galactic plane. As shown in Fig. 3.2, the **Orion Molecular Cloud** (OMC) complex, at a distance of roughly 450 pc, is about 120 pc long. The total mass of gas is estimated to be a few $10^5\, M_\odot$. It can be divided into three clouds, Orion A, Orion B (containing NGC 2024) and Mon R2, each consisting of a few thousand solar masses.

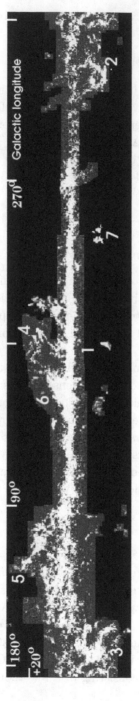

Fig. 3.1 Molecular clouds in the Milky Way: a composite picture from a CO survey of the Galaxy. In galactic coordinates, the (1) Galactic centre lies near 0° longitude. Also marked are (2) Orion, (3) Taurus, (4) Ophiuchus, (5) Polaris Flare, (6) Aquilla Rift and (7) Chamaeleon (Credit: T. M. Dame and P. Thaddeus, Harvard-Smithsonian Center for Astrophysics).

The clouds appear connected to the Galactic plane by two remarkably long slender filaments.

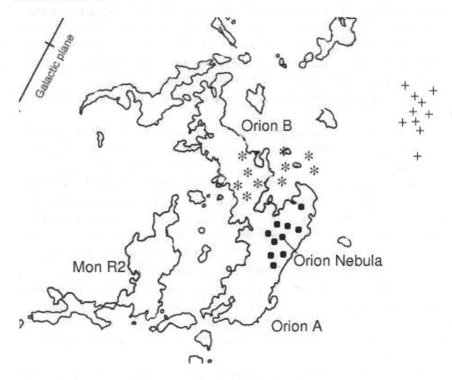

Fig. 3.2 The Orion Molecular Clouds (contours) and OB associations (symbolic) (in equatorial coordinates – as observed – south of the Galactic plane). The oldest stars are in the subgroup OB1a (plus signs, 12 Myr old). The age sequence continues: subgroup OB1b (stars, 6 Myr) and subgroup OB1c (squaress, 2–5 Myr). The youngest stars are densely packed in the Orion Nebula which contains the brilliant Trapezium stars (< 2 Myr old). (Credit: CO data extracted from R. Maddalena et al. 1986, ApJ 303, 375).

The evolution of star formation in large clouds is exemplified by the Orion complex. Massive stars, with the spectral classes O and B assigned to them, have been forming here within four distinct groups called OB associations. These contain extremely bright and hot stars with mass exceeding about $10 M_\odot$.

One generation of massive stars in region 1a formed 12 Myr ago. The parental cloud is absent, having been completely disrupted or transformed into stars. The stars are now free from molecular material and not obscured. Although still clustered, they slowly drift apart. A second generation of stars has been forming for about 7 Myr just on the right edge of the cloud, region 1b, sweeping up and disrupting the cloud in the process. Some stars are now visible while others remain embedded. The latest generation only began about 3 Myr in location 1c. All these objects are still embedded and star formation is ongoing. Indeed, there are many cores which appear to be just on the verge of collapse. This region contains strings of many very young protostars which react with the cloud through powerful jets and winds of gas, driving shock waves and inflating lobes of fast-outflowing molecular gas.

Massive bright stars are forming in the south, in the Trapezium region, appropriately called because of the placement of the four brightest young stars. The bright stars light up the cloud material nearby producing the magnificent Orion Nebula which we see in reflected light. In addition, molecular gas is photo-dissociated by the far-ultraviolet radiation from the OB stars, producing very strong infrared emission in the fine-structure lines of oxygen and carbon. Although the massive stars are prominent and appear to re-structure the complex, giving it the V-shape, lower mass stars are present in great numbers throughout the clouds.

Just behind the Orion Nebula H II region lies the Orion-KL region which contains the densest core in the cloud and a cluster of luminous infrared stars with a total luminosity of $10^5 \, L_\odot$. The core mass is $200 \, M_\odot$ contained within just 0.1 pc. Consequently, this very young region has been the focus of many studies.

Other regions in which stars of high mass are forming are nearby to Orion: the Rosetta Cloud and Gem OB1 complexes. As noted above, however, most of the molecular gas lies in the inner Galaxy. The Sagittarius B2 cloud is massive, $3 \times 10^6 \, M_\odot$, and just 200 pc away from the Galactic centre. It contains small-scale structure, hot cores and outflows. Extending away from the Galactic centre is a thin intense ridge of CO emission. This is associated with a so-called molecular ring, probably consisting of several spiral arms, located a few parsecs from the Galactic centre.

3.3.3 *The Eagle*

The Eagle Nebula shown in Fig. 2.2 is a very young star formation region containing dark columns of cold molecular gas and dust dubbed the 'Pillars of Creation' or 'elephant trunks'. Many thousands of deeply-embedded young stellar objects are detected in near-infrared images. In addition, 73 small evaporating gaseous globules (EGGs) have been detected on the surface of the pillars. Two of the pillars have relatively massive young stars in their tips. However, most of the EGGs appear to be empty: about 15% show evidence for young low-mass stars or brown dwarfs. There are also a large number of X-ray sources. These are more evolved young stars not associated with the evaporating gaseous globules (EGGs) as will be discussed in §9.7. They suggest that star formation has been proceeding for millions of years.

It appears that this structure was formed by the interaction of interstellar clouds with multiple supernova explosions and stellar winds from massive stars near the Galactic plane. These act to compress the clumps which become exposed, thus triggering star formation. Although the Orion Nebula and the Eagle Nebula appear disimilar, this is largely attributed to the orientation: we observe the Orion Nebula face on and so observe through it (see Fig. 7.2 whereas the Eagle Nebula is clealy edge on to our line of sight.

3.3.4 *Breeding grounds for low-mass stars*

The **Taurus-Auriga-Perseus Cloud Complex** is the most intensively studied star-forming region containing only low mass stars. It is a dark, quiescent complex located towards the Galactic anti-centre. In contrast to the regions described above, the cloud complex has long been regarded as a site where star formation occurs spontaneously. The total complex mass is $\sim 2 \times 10^5\,M_\odot$. Estimated masses of fragments cover a wide range with clumps embedded in more massive envelopes, all the way down to a few solar masses. The Perseus molecular cloud is located at a distance estimated to range from 200 to 350 pc and has a projected area of dimension $10\,\mathrm{pc} \times 30\,\mathrm{pc}$ with an estimated mass of $10^4\,M_\odot$.

The Taurus Cloud (TMC) itself (Fig. 3.3) lies relatively close-by, at a distance of 130–160 pc. Its gas mass is $\sim 3 \times 10^4\,M_\odot$ and size 30 pc. The most striking aspect is its striated morphology, displayed in Fig. 3.3. About 150 young stars, mainly T Tauri stars, are in the T association (§12.4) weakly clustered within individual clumps embedded in the striations.

There is evidence that each star in the TMC has developed in an isolated environment. Firstly, the potential trigger in the vicinity, the Cas-Tau OB association, is already old (25 Myr) and the stars are widely dispersed. Secondly, the number density of the young stellar objects (YSOs) in the TMC is quite low, 10–20 stars pc^{-3}, implying that neighbouring systems do not generally gravitationally interact with each other during star formation.

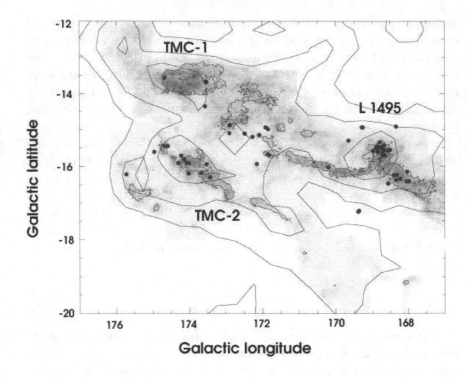

Fig. 3.3 The Taurus Molecular Cloud, combining data from CO isotopes including ^{13}CO (grayscale) and C^{18}O (J=1–0) (inner contours). The locations of only the youngest T Tauri stars are shown (dots) (Credit: adapted from F. Palla & S.W. Stahler, 2002, ApJ 581, 1194).

On the other hand, some evidence for a more systematic star formation mechanism is derived from a chemical analysis. The TMC-1 component is a prominent example of chemical inhomogeneity. The south-eastern part of the cloud is abundant in 'carbon-chain' molecules such as CCS and cyanopolynes (CPs), while the north-western part is abundant in molecules such as NH$_3$ and N$_2$H$^+$. Since carbon-chain molecules are considered to be abundant in the early stage of chemical evolution and the abundances of NH$_3$ and N$_2$H$^+$ are enhanced in the later stage, this separation has been interpreted to be consequence of sequential chemical evolution along the cloud.

More evidence is derived from the numbers and ages of the T Tauri stars and the lack of post T Tauri stars. This implies that, in the past, star formation proceeded at a low rate until about 4×10^6 yr ago. There then followed a rapid rise toward the present epoch. This could not have plausibly occurred spontaneously over 30 pc but, instead indicates that a triggering wave crossed the cloud at a speed of ~ 10 km s^{-1}.

Lupus and ρ Ophiuchus contain similar clouds to the TMC. They are situated above the plane towards the Galactic centre. The ρ Ophiuchus cloud at a distance of 160 pc contains a total mass of molecular gas of $\sim 3 \times 10^3$ M$_\odot$. It contains a much denser concentration of gas, the L1688 cloud, which has allowed a detailed study of the distribution of gas and embedded protostars. Visually, it is a cobweb: the cloud filaments resemble strands in a spider's web. Physically, several structures possess head-tail or cometary morphologies, as though blown and shaped by the impact of a wind. The filamentary tails can be either cold un-shocked ambient material or material streaming away from cores. There is ample evidence, including spatial variations in the chemistry, that the star formation has been triggered by winds and supernova blast waves arriving from the nearby Sco OB2 association as well as nearby expanding H II regions.

3.3.5 *Isolated clouds*

Bok globules were discovered as small isolated clouds. They were not identified through emission but as dark patches on the background curtain of stars which covers the Milky Way. At visible wavelengths, these patches are caused by the presence of a high column of obscuring dust. They are examples of 'Dark Clouds' which can also appear as dark patches superimposed on bright nebula (Fig. 3.4). With their discovery, Bart Bok suspected that these were sites where we might uncover the first signs of star formation. Indeed, this has proven correct with over half of them possessing protostars or young stars. While most are low mass, containing a few solar masses of cloud material, others contain several hundred solar masses and may form generations of stars over a few million years.

Considerable attention is now being given to the globules since we hope to isolate the formation mechanism in the absence of external triggers. They are typically 0.1–1 pc in size with quite rough distorted shapes.

An example of a quiescent Bok globule is B 68 shown in Fig. 3.4. There is no evidence for associated protostars or for collapse towards a protostar. It possesses a small mass and a radial density distribution consistent with a static and constant temperature cloud (i.e. resembling a Bonnor-Ebert sphere as described in §4.3.3). It is, however, far from spherical and contains peculiar chemical distributions and internal motions not consistent with a

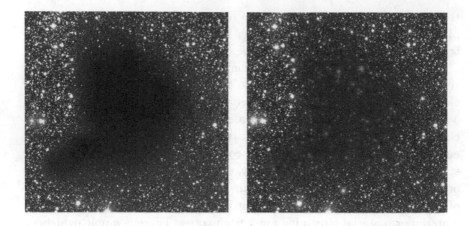

Fig. 3.4   The Bok globule B 68: composite of visible and near-infrared images of Barnard 68 obtained with the 8.2-m VLT. At visible wavelengths (left), the small cloud is completely opaque. Including near-infrared emission (right), some background stars are detected. (Credit: J. Alves, FORS Team, VLT, ESO)

static configuration.

Other Bok globules such as B 335 and CB 34 are actively forming stars. The CB 34 globule is observable in emission at submillimetre wavelengths, where the central region is found to be clumped. The clumps contain protostars, observed in the near-infrared. The protostars drive gas outflows in the form of bipolar outflows and chains of bullet-like knots. Spreading out from the clumps are young stars which suggest that the globule is in an advanced stage but still only a few million years old.

Not all clouds will form stars. Examples are High Latitude Clouds (HLCs) and Chaff. The HLCs are molecular, and, as their name suggests, not found in the Galactic plane. They possess high turbulent pressure but are of too low a mass to be self-gravitating. Chaff is the name given to the equivalent of HLCs but in the Galactic plane. These consists of low column density material detected in the vicinity of GMCs and the spiral arms with masses under $1000\,M_\odot$. The total mass in Chaff in our Galaxy is estimated to be $7.5\text{--}15 \times 10^7\,M_\odot$.

3.4   The Internal Dynamics and Structure of Clouds

The masses of the stars which eventually form could be pre-set by the masses of the clumps and cores. Therefore, astronomers have been very creative in their attempts to characterise cloud structure. Probability distribution functions (PDFs) for density or velocity provide a robust basis for the analysis of cloud structure. There are various more sophisticated techniques to describe the volume distribution, the sky distribution and the velocity distribution. The latter requires spectroscopic methods. That is, we obtain images of the cloud in specific narrow wavelength channels, which correspond to channels in radial velocity (the velocity component in the line-of-sight). This is called a clump decomposition technique, which uses the relative shift in radial velocity to distinguish cloud components superimposed on the sky. We can then derive some information in the third spatial dimension.

The results are crucial to *how* a cloud evolves into stars, *how much* of a cloud evolves into stars and their *distribution* in space and in mass. The observed masses of clouds generally follow a power-law distribution with the number of clouds per unit interval in mass given by $dN_{cloud}/dM \sim M^{\alpha_i}$. It is convenient to employ a mass-weighted distribution:

$$M \frac{dN_c}{dM} = \frac{dN_c}{d\log M} \propto M^{\alpha_c} \qquad (3.7)$$

since then a positive index α_c means that most clouds are massive and a negative index means that most clouds are low mass.

We actually find that most clouds are low mass with the index almost invariably in the range $\alpha_c = -0.6 \pm 0.2$. For a large sample of clouds in the inner Galaxy $\alpha_c = -0.6$, and in the Perseus Arm $\alpha_c = -0.75$. The power-law holds over a wide range of masses from GMCs down to the high-latitude clouds with mass $10^{-4}\,M_\odot$.

It is highly significant that most of the clouds are small but most of the mass is contained in the most massive clouds (since $1 + \alpha_c$ is positive). It is equally important that we do not observe any giant molecular clouds in our Galaxy above a certain limit. In fact, the power-law possesses a sharp cut-off at $6 \times 10^6\,M_\odot$. Writing the number distribution of clouds within the solar circle in the form

$$M \frac{dN_c}{dM} = 63 \left(\frac{M}{6 \times 10^6\,M_\odot} \right)^{-0.6}, \qquad (3.8)$$

one can immediately see that the cut-off is very sharp – we would expect about 100 extreme GMCs above the break if we extrapolate the power law. Instead, there are actually none.

In 1981, Larson brought together several significant results which have remained essentially unchallenged. He uncovered correlations between the cloud densities, cloud sizes and the internal velocities within a cloud. It first became clear that the motions could only be described by supersonic turbulence. The line widths were non-thermal, far exceeding the thermal velocity dispersion v_{th} which is the line widening attributable to the gas temperature T. The sound speed in an isothermal cloud would be

$$c_s = 0.44\mu_m^{-1}\left(\frac{T}{10\,\mathrm{K}}\right)^{1/2}\,\mathrm{km\,s}^{-1} = 0.19\left(\frac{T}{10\,\mathrm{K}}\right)^{1/2}\,\mathrm{km\,s}^{-1}, \qquad (3.9)$$

where μ_m is the mean molecular weight. This is equivalent to the thermal velocity dispersion. The bulk of the clouds may often be treated as isothermal, exhibiting only small temperature variations. In contrast to the low sound speed, observed velocity dispersions are large, regularly exceeding $1\,\mathrm{km\,s}^{-1}$.

Furthermore, for many clouds, the velocity dispersion (or line width $\Delta v = 2.355\,\sigma_v$) is related to the cloud size expressed in parsecs by

$$\sigma_v = (0.72 \pm 0.07)R_{pc}^{0.5\pm0.05}\,\mathrm{km\,s}^{-1}. \qquad (3.10)$$

There are numerous exceptions reported, but it is probably safe at present to expect indices in the range 0.5 ± 0.2. This cornerstone relationship holds from cloud to cloud as well as within individual clouds.

A similar relationship exists for low-mass cores. For high-mass cores, however, the relationship for non-turbulent motions is significantly shallower:

$$\sigma_{nt} \approx 0.77 \ R_{pc}^{0.21}\,\mathrm{km\,s}^{-1}, \qquad (3.11)$$

suggesting some support or feedback from stars forming within.

The second major result is that the density is related to the cloud size by $\rho_c \propto R_c^{-1.1}$. For GMCs inside the solar circle this can be more accurately expressed in terms of the average column of hydrogen nuclei, N(H),

$$\frac{N(H)}{10^{22}\,\mathrm{cm}^{-2}} = (1.5 \pm 0.3) \ R_{pc}^{0.0\pm0.1}, \qquad (3.12)$$

which is the product of the average density and the cloud depth. When written in terms of cloud mass, we find $M \propto R^2$ as a rough guide. This was initially taken as strong evidence that the clouds are stable and in virial equilibrium since the above relations yield $\sigma_v^2 R/M$ to be a constant, on putting $M \propto N(H) R^2$, which satisfies the virial formula, Equation 3.3.

The above relationship, however, does not extend to low mass clouds and clumps for which $M_c \propto R_c^{1.8}$ is more appropriate, giving

$$\frac{N(H)}{10^{22}\,\text{cm}^{-2}} = (1.5 \pm 0.3)\,R_{pc}^{-0.2\pm0.1}. \tag{3.13}$$

To determine if a cloud could be bound we evaluate the *virial parameter* α:

$$\alpha = \frac{5\sigma^2 R_c}{G\,M_c}. \tag{3.14}$$

Putting the formula to define virial mass (Eq. 3.4) in the form

$$M_{vir} = 209 \left(\frac{R_c}{1\,\text{pc}}\right)\left(\frac{\Delta v}{\text{km s}^{-1}}\right)^2 M_\odot \tag{3.15}$$

and determining the cloud mass M_c from CO observations, yields power-law fits

$$\frac{M_{vir}}{M_c} = 0.85 \left(\frac{M_c}{10^4 M_\odot}\right)^{-0.38\pm0.03} \tag{3.16}$$

for the Cepheus-Cassiopeia region and similar relationships for other regions.

The result is that we find most GMCs are indeed gravitationally bound with $\alpha \sim 1$. The virial formula applies accurately for $M_c > 10^4{}_\odot$ but similar scaling relations are also now extended to include gravitationally unbound lower mass clumps in many regions, such as the Gemini-Auriga, Cepheus-Cassiopeia and Cygnus complexes. A lot of the lower mass clouds are unbound with α typically ~ 5. Also, many clumps, such as those in the Rosetta cloud, are not gravitationally bound. The typical value for the clumps in ρ Ophiuchus is $\alpha \sim 10$. These clumps must be either confined by an external pressure or in the process of expanding. On the other hand, dense cores are necessarily bound and could well go on to form single stars or small systems such as binaries.

The lack of observed clouds or sub-components with $\alpha \ll 1$ is of great significance. Such clouds would be unstable and collapse abruptly in the absence of any other support mechanism. This indicates that there is indeed no other support mechanism, such as the pressure of a the magnetic field, which could become totally in command.

The σ–R law and the lack of virial equilibrium could be the consequences of turbulence. A turbulent energy cascade, if it were to follow a Kolmogorov-type law for incompressible gas, would predict $\sigma_v \propto R^{1/3}$, slightly shallower than observed. The origin of this expression can be seen from a dimensional analysis of a turbulent cascade. In what is termed the 'inertial range' of scales L and velocity amplitude U, the rate at which turbulent energy is transferred to smaller scales is proportional to the energy $MU^2/2$ and the inverse time scale U/L. Hence, in a steady energy cascade we find the rate of flow of energy MU^3/L is a constant. Therefore, U $\propto$ $L^{1/3}$ could be expected. This scenario is fully explored in §4.4.

The ρ Ophiuchi molecular complex yields distinct properties with $\alpha_c = -1.1$ and $N(H) \propto R^1$ for the clumps with masses between 3 and 300 $M_\odot$. The latter implies that the density and size are uncorrelated. Virial equilibrium does not apply with σ_v nearly independent of R; the low-mass clumps deviate most from virial equilibrium.

We note here that pre-stellar cores, which are dense cores thought to be on the verge of collapse, have also been analysed in ρ Ophiuchus (see §6.2). These display a number distribution with $\alpha_c = -1.5$ below $\sim 0.5\,M_\odot$, steepening to $\alpha_c = -2.5$ above. This, as we shall see, is reminiscent of the Initial Mass Function of the stars, suggesting a direct evolution of these discrete cores into stars.

3.5   Rotation Trends

Not much has been known about cloud rotation apart from the fact that they don't rotate much. This requires explanation since rotation at high speed is expected according to the principle of angular momentum conservation. Clouds should spin up as they contract (see §7.6).

As an indicator of spin, we can measure the spatial variation of the radial component of velocity across a cloud. As a tracer, we employ the CO molecular lines at millimetric wavelengths. A velocity gradient is then interpreted as rotation although that may not always be the case. For example, dynamical shearing between a cloud and the ambient medium, chance superposition, wind impacts from supernova and outflows from young stars may all lead to radial velocity gradients. Systematic extended gradients are, however, attributed to rotation and these require explanation. In particular, a linear shift of velocity in a spectral line with a spatial coordinate is predicted for a cloud uniformly rotating.

The first result is that the spin axes are not randomly orientated on the largest cloud scales. Systematic gradients in velocity extend in some cases over tens of parsecs. If clouds in our Galaxy derive their spin directly from the shear on galactic scales, then we would expect the spin axes to be aligned with that of the Galaxy. This is, in fact, what is found with approximately equal numbers orientated towards the north and south galactic poles.

The smaller clouds and cores, however, do not follow the trend, consistent with random orientations over all directions. This indicates that they do not derive their present angular momentum from the galactic shear but turbulent processes or interactions between fragments are responsible. A further highly relevant fact is that the directions of cloud rotation and elongation are not statistically related.

The rate of rotation in radians per second, Ω, is correlated with the cloud size, R. Smaller clouds rotate faster, in general. In theory, conservation of cloud angular momentum, J_c, for a cloud of mass M_c, size R_c and rotation speed v_{rot} would predict that the product of rotation speed and radius remain constant:

$$J_c \propto M_c v_{rot} R_c \propto M_c \Omega R_c^2. \tag{3.17}$$

This implies $\Omega \propto R_c^{-2}$. The observed trend, however, is remarkably shallower with $\Omega \propto R_c^{-0.6}$:

$$\frac{\Omega}{10^{-14}\ \text{s}^{-1}} = 10^{0.34} \left(\frac{R}{1\ \text{pc}} \right)^{-0.56}. \tag{3.18}$$

Hence, although smaller clouds rotate faster, the increase is not consistent with angular momentum conservation.

This is succinctly expressed in terms of the angular momentum per unit mass, J/M, called the specific angular momentum. For any individual core, the radial velocity gradient may not be a good indicator of its specific angular momentum. However, the distribution of deduced angular momenta from a large sample of cores with different random velocity fields is close to the distribution of actual angular momenta of these model cores if one assumes $J/M = p\,\Omega R^2$. For centrally condensed cores, the standard value of p = 0.4 overestimates the mean intrinsic angular momentum by a factor of 3. On measuring according to $J/M = 0.4\ \Omega\ R^2$, we find that $J/M \propto R^{1.4}$. This suggests that, in most scenarios, the smallest clouds and condensations have lost their angular momentum during their contraction. Much research has been expended in deriving explanations for this fact with the process of magnetic braking being best explored (see §7.6).

The tightest correlation is between J and M. This is, in part, because both these quantities have been determined over many orders of magnitude. Cloud masses range from 0.1 to 10,000 M$_\odot$ and angular momentum from

10^{-4} to 10^6 in the units of $M_\odot$ km s^{-1} pc (see Table 3.5). The relationship in this measurement unit takes the form

$$J = 10^{-2.22} \left(\frac{M}{M_\odot} \right)^{1.68}. \tag{3.19}$$

This is fully consistent with the above formula for J/M combined with $M \propto R^2$, as shown above from Eq. 3.12. The tightness of this correlation, however, implies that the loss of angular momentum is related closer to the decrease in mass rather than the contraction. This is thought to be indicative of the physical mechanism.

Since we expect the specific angular momentum to be conserved during a contraction, it is useful to tabulate the typical values for J and J/M. It is then immediately clear from the final columns of Table 3.5 that angular momentum is not just sub-divided into the fragments but is transferred around, being extracted from the smaller scale structures. In fact, in the absence of this extraction, the centrifugal force would soon prohibit the contraction (see §7.6).

Table 3.3   A summary of the rotational properties of star formation structures and objects. Note that the angular momentum per unit mass J/M decreases by many magnitudes as the scale decreases.

Structure	Ω s^{-1}	J ($M_\odot$ cm^2 s^{-1})	J/M (cm^2 s^{-1})
Giant MC	3×10^{-15}	3×10^{29}	3×10^{25}
Molecular clump/cloud	0.3–3×10^{-14}	10^{22}–10^{27}	10^{23}
Molecular core	1–10×10^{-13}	3×10^{21}	0.4–30×10^{21}
Wide Binary[a]	2×10^{-11}	10^{21}	10^{21}
Circumstellar disk [c]	2×10^{-10}	5×10^{18}	5×10^{20}
Planet (Jupiter orbit)	2×10^{-8}	10^{17}	10^{20}
Close Binary[b]	6×10^{-6}	10^{19}	10^{19}
T Tauri star	2×10^{-5}	5×10^{17}	5×10^{17}
Present Sun	2×10^{-6}	10^{15}	10^{15}

[a] 10^4 year period

[b] 10 day period

[c] 100 AU, 0.01 $M_\odot$ disk

3.6 Structural Analysis: Fractals and Filaments

The GMCs and large clouds are highly clumped, as evident from the CO maps. To quantify this, we combine an observed GMC mass and a typical extinction-derived column density into the predicted mean density:

$$n(H) = 84 \left(\frac{M}{10^6 \, M_\odot} \right)^{0.5} \left(\frac{N(H)}{1.5 \times 10^{22} \, \text{cm}^{-2}} \right)^{-3/2} \text{cm}^{-3}. \qquad (3.20)$$

In contrast, we infer densities exceeding 1000 cm^{-3} from the column densities of the defined clouds and clumps. Hence, volume filling factors are typically under ten per cent. Surprisingly, we still have little knowledge of the inter-clump medium in a molecular cloud which could be atomic or molecular.

A fractal or hierarchical description of clouds is suggested by the lack of a scale in the power-law mass distributions of clumps. We have seen that the gas is indeed highly textured. The 'self-similarity' is present on scales stretching from hundreds of parsecs down to a tenth of a parsec, from spiral arms down to molecular clumps. Furthermore, cloud boundaries show a complex structure and have highly irregular shapes that do not correspond to equilibrium configurations. Rather, they resemble the amorphous terrestrial clouds, which are known to be transient dispersing structures.

The fractal dimension of a cloud perimeter, D, is found by relating the perimeter length to the cloud area, $P \propto A^{D/2}$, so that simple, rounded shapes correspond to $D = 1$. Analysis of observed clouds yields a projected fractal dimension $D = 1.3 \pm 0.3$ in several cases. This value is intriguingly similar to fractal dimensions measured in turbulent flows.

It is not yet clear how much interstellar structure is fractal. Quite commonly observed regular features are filaments and shells, detected in dust and molecular tracers. Filaments can extend several parsecs and are marked by a string of embedded starless and protostellar cores. These elongated structures are not generally aligned with rotation or magnetic field axes. Filaments can be remarkably straight, but there are also examples of curved and wavy forms. Spectacular long, twisted structures were observed by the Hubble Space Telescope in the Lagoon nebula. Elsewhere, a system of helix-like, interwoven dark filaments were found in the Rosetta nebula.

3.7   Summary: Invisible Clouds

We have recorded many correlations between cloud properies which need explaining. All this knowledge arises from the clouds we observe. Are these the clouds which actually form the stars or are they the clouds which, in this state, make themselves obvious? In our snapshot of the Universe, clouds which have a fleeting existence would not be fairly represented. Yet, we cannot *sweep search* a volume of our Galaxy. We could also be overlooking a population of invisible clouds, too cold to detect in molecular emission lines or dust emission. There may also be a population of clouds too diffuse to distinguish from the background. To the opposite extreme, a large collection of compact clouds of high density and very low mass have also been proposed as a means to hide dark matter within the interstellar medium.

Such claims are not easy to confirm or dismiss until decisive observations are made. Nevertheless, we proceed in the knowledge that the molecular clouds we observe contain protostars and are surrounded by young stars. We also note that the young stars we observe are almost exclusively associated with molecular clouds. Exceptionally, we detect young stars without any obvious parent cloud and we then speculate that the star has been orphaned either by rapid cloud dispersal or as a runaway.

Chapter 4

Cloud Formation, Evolution and Destruction

We now begin to trace the journey towards a star. How long does this take? The answer is surprisingly short: a good many clouds already contain new stars and these stars tend to be young. The typical cloud cannot spend long, if any time at all, in a dormant state.

The first challenge is to accumulate gas into a molecular cloud. One dilemma is to decide what occurred first: the production of the cloud or the molecules? We all agree that a star-forming cloud must become gravitationally bound. Here, we shall investigate what this implies and what bound states are possible. To do this, we need to bring in concepts from hydrostatics and hydrodynamics. Changes in pressure and gradients in pressure take on great significance. Where will the environment's pressure confine a cloud? And where will a cloud lose pressure, deflating like a balloon? Or, where will the cloud's own gravity provide the confinement?

Theory is vital in this chapter since the changes in clouds cannot be observed on human time scales: the evolution is too slow and must be hypothesised. Many scenarios begin from a static equilibrium. This provides the initial conditions to which each protostars can be traced back. Yet the observed motions are impressively fast and we find that classical theory cannot keep up. We are led towards something more violent.

4.1 The Ages of Clouds

Age estimates for a typical Giant Molecular Cloud are discordant, lying in the range between 10^9 years and 4×10^6 years. Surely we can be a bit more precise? The higher value is derived on global grounds, simply from the total mass available, $3 \times 10^9 \, M_\odot$, and a star formation rate in the Galaxy estimated to be $3 \, M_\odot \, yr^{-1}$. Accordingly, clouds were thought to be senile and in need of support.

Secondly, star formation takes place primarily in the spiral arms where molecular clouds are either swept together or atomic clouds are compressed into clouds. Spiral arms are caused by density waves: the effects of gravity in a differentially rotating disk of material. The spread of clouds in the arms would indicate an age of order 10^8 years. This argument is also quite weak since it is not clear that clouds form exclusively in the arms. However, in most galaxies with distinct spiral arms, molecular clouds are generally confined to the arms.

A third argument employs the time scale for the production of OB stars. These are the massive stars such as formed in the Orion complex (see §3.3.1). Their radiation and winds should be capable of ripping apart a cloud within about 3×10^7 years. Even more constraining, in those regions with stars older than about 5×10^6 years there is no associated molecular gas.

Finally, the minimum time is clearly the free-fall time. This is the average time taken under gravitational acceleration for a part of a cloud to collapse unimpeded under its own gravitational pull starting from rest (by unimpeded it is meant that the cloud is devoid of all pressure support). Undertaking the calculation for a uniform cloud, we find that the total collapse time is independent of where the collapse started from and the cloud remains uniform. This yields the time

$$t_{ff} = \sqrt{\frac{3\pi}{32G\rho}}, \tag{4.1}$$

at which the density reaches infinity. This implies the collapse would last just $t_{ff} = 44 \times 10^6 / n^{0.5}$ years $= 4.4 \times 10^6$ years for an average GMC density of $100 \, \text{cm}^{-3}$ (see Eq. 3.20). One can also define a dynamical time scale R_c/σ in terms of the dispersion in velocities

$$t_{dyn} = 0.98 \times 10^6 \left[\frac{R_c}{1\text{pc}} \right] \left[\frac{\sigma_v}{1 \ \text{km s}^{-1}} \right]^{-1} \quad \text{yr.} \tag{4.2}$$

This suggests that the observed velocities should result in a rapidly changing cloud morphology, whatever the density.

Current evidence is coming down on the side of the shorter time scales and this is changing our view on how clouds evolve. Instead of being slowly collapsing or supported objects, molecular clouds are dynamic and ephemeral, resembling terrestrial clouds. The different layers of evidence converge towards a single conclusion: the typical cloud lifetime is just a few free-fall or dynamical time scales.

For many years, astronomers attempted to find means of support for the proposed senile cloud, in order to impede their immediate gravitational collapse. The obvious candidate was actually observed: turbulence. The width of observed emission lines implies turbulent motions within clouds. Hence, it seemed obvious to some that turbulence hinders cloud collapse and is the reason why clouds still exist.

It became increasingly evident, however, that we were struggling with this explanation. First, the origin of the turbulence was hard to pinpoint. Second, the turbulence needs continual replacement since it dissipates as fast as it is produced. Resourceful astronomers invoked particularly resilient waves called Alfvén waves (§7.2) in which energy can be temporarily stored as magnetic energy, to be re-released back into the turbulent motions. Computer simulations finally laid serious doubt, not on the existence of such waves, but on their durability.

At the same time our knowledge of clouds and the distributions of young stars had been accumulating. The hard reality was that clouds are associated with quite young stars, typically under 10×10^6 years old. This was found to hold for both the subgroups of OB stars as well as for lower mass T Tauri stars associated with dark clouds. Furthermore, very few GMCs and a minority of dark clouds are not associated with young stars. Hence the total life of clouds cannot far exceed 20×10^6 years on average. In addition the isolated dark clouds, the Bok globules, have been estimated to be a maximum of 3×10^6 years old.

The phases of rapid evolution deduced from GMC observations are sketched in Fig. 4.1. At first, only low-mass stars form in a cold dense cloud. The ultraviolet photons do not create large ionised regions and molecular gas survives. At some point of the cloud's history, high-mass stars form. What triggers this change is still not clear. An external factor is probably responsible: distant high-mass stars may be triggering larger fragments of the cloud to collapse. Or, a number of low-mass protostars might coalesce. The newly formed massive stars ionise and dissipate the local cloud, bringing star formation to an end except in more distant suburbs. Therefore, while the local cloud is dispersed, the activity of the massive stars triggers a collapse in distant waiting clouds.

There is also anatomical evidence for rapid cloud evolution: many clouds are chemically immature. Various trace molecules appear not to have had time to reach their equilibrium abundances. Instead, the regions are still

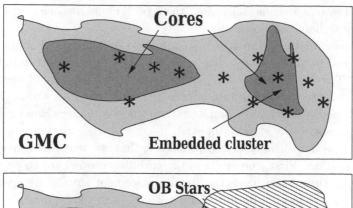

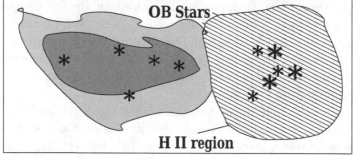

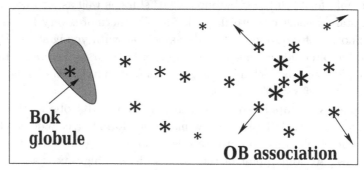

Fig. 4.1 A schematic sequence of events leading to the formation of an OB association and an isolated Bok globule out of a Giant Molecular Cloud. The evolution occurs from top to bottom in 10 Myr. Most of the cloud disperses and the final cluster is not gravitationally bound.

in the course of processing the molecules, suggesting the clouds have only existed for a few million years (see §3.3.4). It has also been found surprising that more molecules within dense cold clouds have not condensed or 'frozen out' onto dust grains as predicted if the clouds were stable static entities.

These enigmas have actually been around for decades. Yet the consensus was that clouds were old, forcing alternative explanations to be sought. For example, the turbulent motions could cycle the material between the interior and the surface where the exposure to ultraviolet radiation could rejuvenate the chemical state.

How do we reconcile the rapid cloud evolution with the relatively low rate of star formation, $3\,M_\odot\,yr^{-1}$, in our Galaxy? Given the molecular mass and the short lifetime, then $150\,M_\odot\,yr^{-1}$ is assembled into clouds. Accordingly, we argue that only 2% of the total cloud material gets converted into stars. The remaining 98% is dispersed!

We define the *efficiency* as the portion of a cloud which is turned into stars before the cloud disperses. A low efficiency means that the cluster of stars which form is unlikely to have sufficient mass to be bound (depending on how tightly bound the cloud is and the relative sizes of the cloud and cluster). The *global* star formation efficiency might be indeed be low but the *local* efficiency of individual clouds is estimated to be extremely high. In cloud complexes such as Orion, star formation is confined to specific small locations. Typically, 10–40% of a cloud core is converted into a cluster of stars. Such high efficiencies mean that little residual gas is left to be dispersed and a gravitationally bound cluster is likely.

We may conclude that there is overwhelming evidence in favour of transient clouds. Some fraction of a cloud collapses while most of a cloud is recycled. Clouds are dynamic: a quasi-equilibrium configuration in which gravity is held up by turbulence is not necessary.

4.2 The Origins of Clouds

4.2.1 *Formation of the giant clouds*

There are two types and two states of clouds. Clouds can be either diffuse or self-gravitating, atomic or molecular. These are independent conditions: even diffuse clouds are large enough for their interiors to become protected from UV light, allowing molecules to form. High Latitude Clouds (HLCs) are good examples. HLCs are molecular, and, as their name suggests, not found in the Galactic plane. They possess high turbulent pressure but are of too low a mass to be self-gravitating.

Hence, four possible cloud conditions exist. Which one occurs will depend on the site conditions, including the local UV radiation flux and the dust abundance.

We still have to determine how and where the atoms combine into molecules. This may proceed within the diffuse clouds which are subsequently compressed into observable structures. Or, diffuse atomic clouds may first be compressed into dense clouds where molecules can form efficiently.

Mechanisms which have received support for directly forming molecular clouds are as follows. In practice, each mechanism may instigate one or more of the other mechanisms and so act indirectly.

1. Agglomeration. Clouds grow by random collisions and coalescence of diffuse molecular clouds (pre-formed from atomic clouds). The process is enhanced during the passage of a spiral arms where orbits are focussed. This is the 'bottom-up' approach.

2. Gravitational instability. Large objects form first and fragment (a 'top-down scenario').

3. Collision within high-speed turbulent flows. Molecular clouds are the tips of the icebergs in the general turbulent flow: molecules form in regions of elevated density, especially after the collision of high density regions.

4. Shock accumulation. Mass is swept up into thick *supershells* by winds and supernova. The supershells eventually become gravitationally bound and fragment.

5. Streaming along magnetic fields. Large-scale fields bend and twist out and through the galactic plane since they are buoyant. Material pours down the field to magnetic 'valleys', locations of low gravitational potential.

6. Thermal instability. Regions of atomic diffuse gas which are slightly cooler but denser than their surroundings may cool faster, losing thermal support. Becoming even denser and cooler, radiation cooling rates may further increase, leading to a runaway process in which flattened clouds form.

The collision or random coagulation hypothesis has been around for a long time in diverse forms. It is now attacked on several grounds. First, the process is slow, taking over 100 Myr to construct a GMC. Second, there is no guarantee that a collision leads to a merger rather than to destruction. Thirdly, the clumps and clouds that we once perceived as distinct ballistic objects turn out to be part of an interconnected network.

Most conclusively, collision models are inconsistent with the observations: there is an insufficient number of small molecular clouds in existence to be able to form the GMCs. Most of the molecular mass is already in the GMCs (§3.4). It may be possible that a vast sea of small molecular clouds has been overlooked. They could be too cold to be detected in the CO tracer. Alternatively, they may be in the diffuse form termed 'Chaff', the equivalent of High Latitude Clouds but in the Galactic plane. The total mass in Chaff in our Galaxy is estimated to be less than 5% of the that in GMCs (§3.3.5). That implies that to be responsible for GMCs, the molecular gas must spend only 10% of its time in the form of Chaff. This is untenable given the already short GMC lifetimes.

Hence it appears that the GMCs must have formed by condensation from atomic material in a 'top-down' picture. This has considerable observational support. First, dense neutral atomic envelopes to molecular clouds are quite common. In addition, some clouds in the spiral arms contain over $10^7 \, M_\odot$ of neutral atomic hydrogen. These so-called 'superclouds' often display sub-structure recognisable as GMCs. The smaller atomic envelopes could be in transit in either direction: dissociated molecular gas or atomic gas in the process of cooling and condensing onto the molecular cloud. It should be remarked that atmospheric clouds form similarly with water droplets condensing in certain regions where air rises, rather than being gathered together after droplet formation.

In contrast, near our Galactic centre, the molecular gas is dominant. This is not surprising since the site conditions are critical. The rate at which molecules form in these dense regions is much higher. Therefore, atoms may only survive for considerable periods in exposed sheets or tiny knots. On some days the cloud cover is thick.

The question as to what large-scale process gathered the material into clouds still remains open. Gravitational instability on a large scale provides a plausible explanation. Here, the interstellar gas is collected directly into large cloud complexes through the growth of sheared density perturbations. Density waves, which clearly collect up the gas and amplify star formation, lead to spiral arms and bars in galaxies. Gravitational instability, or 'swing amplification', may more generally lead to star formation in complex environments.

In a medium as turbulent as the ISM, however, the formation of gravitationally bound clouds hardly requires an instability. The fact that gravitational instability may be present suggests that cloud formation would be enhanced. Hence, our third explanation is the most convincing, with clouds resulting from the large dense regions within a fully turbulent flow. The turbulence and density in these regions are enhanced in spiral arms. The question of responsibility is then restated: what causes or drives the turbulence? We will explore the solutions in §4.4.

Star formation in the extended solar neighbourhood is also believed to have been triggered by the accumulation of gas within a spiral arm into a supercloud, which then fragmented into molecular clouds. The process was initiated some 60 Myr ago when, it is estimated, we would have been part of the Carina arm. An OB association disrupted a parental cloud blowing a gigantic bubble, pushing away a massive ring, called Lindblad's ring. This ring fragmented about 20 Myr ago to form our local GMCs such as Orion and Perseus. Eventually, new clouds were formed in Taurus and Ophiuchus. Now only remnants of the original supercloud can be recognised.

Amongst the alternative models, also likely to occur in some locations is the fourth alternative. Compressive but non-disruptive shock waves are widespread in galaxies. The blast waves from supernovae weaken as the hot cavity they produce expands. Our observationally-derived view of the interstellar medium as 'frothy', in which supernovae and H II regions sweep up shells of gas, favours the supershell mechanism for GMC formation. Galactic shocks may have other causes related to any steep pressure gradient in the interstellar medium. These shocks undoubtedly trigger the formation of stars in existing clouds, as will be discussed, but they may also be capable of compressing gas towards the critical mass where self-gravity takes over. One can envision the driven supershells as producing second generation clouds, having a moderate gathering capacity as compared to the spiral arms. Second or third generations of stars then follow, probably also accounting for our Sun.

Models constructed upon thermal instabilities and other large-scale instabilities can be discarded since they all require some disturbance to a steady state. A steady state rarely exists in the ISM although the hot diffuse gas stands the best chance of reaching such a state. Hence small diffuse clouds could be formed in this manner. The non-linear effects associated with these processes, however, may certainly be responsible for specific cloud structure, taking over command when turbulence or gravity has initiated the cloud formation process. We discuss instabilities further in §4.3.4 below.

4.2.2 *Formation of Bok globules and dark clouds*

Can stars form in isolation, or did they all form in large associations and then drift apart to appear so scattered through the night sky? The presence of young stars in Bok globules suggested that stars could indeed form in isolation. This may, however, prove to be incorrect for the following

reason. Close examination of wide regions of sky indicates that these globules are not so distant from molecular cloud complexes. They may have
originally been part of these cloud complexes but lost contact. Or, they
may have been associated with dense dusty ridges or filaments of atomic
gas extending away from molecular clouds. These ridges may have found
themselves isolated and at arm's length from the UV radiation of OB stars.
Thus, molecules may have begun to form. In turn, the molecules cooled
the ridge which then gravitationally collapsed and fragmented into clumps.
Interclump material with a low column density is left behind which could
be promptly dispersed or ionised.

4.3   The Fate of a Cloud

4.3.1   *Gradual evolution: cloud pressure*

Which clouds will form into stars? How can we decide if a cloud will
collapse or disperse? To tackle these questions, we first need a mathematical
description for the motions of the gas. We describe the gas as a fluid, which
means we do not concern ourselves with the motions of individual molecules
but with their collective motions. This is in general fine since the molecules
do indeed move together, keeping the same neighbours for considerable
time. As we have seen, the cross-sectional area which a molecule effectively
presents to its neighbours is $\sim 10^{-15}\,\mathrm{cm}^2$ (§2.3.2). With a thermal speed
of $0.2\,\mathrm{km\,s^{-1}}$, collisions occur every $1/(\sigma\,\mathrm{n}\,\mathrm{v}_t)$ seconds, or about $1000/n$
years. Since n exceeds $100\,\mathrm{cm}^{-3}$, the collision time is tiny compared to the
cloud evolution and dynamical time scales. Therefore, a fluid treatment is
correct.

The motion of a fluid element of gas is determined by the sum of the
applied forces. The obvious forces on the cloud are due to pressure differences in the gas, gravity and magnetic fields. The force on a unit volume of
fluid of density ρ is written in the form of the fundamental hydrodynamic
equation of motion:

$$\rho\frac{d\mathbf{v}}{dt} = -\nabla\mathrm{P_t} + \rho\,\mathbf{g} + \frac{1}{4\pi}(\nabla \times \mathbf{B}) \times \mathbf{B} \qquad (4.3)$$

where $d\mathbf{v}/dt$ is the acceleration resulting from the gradient in pressure,
$\nabla\mathrm{P_t}$, the total gravitational acceleration on unit mass, $\mathbf{g}$ and the magnetic field B-term, called the Lorentz force. We postpone discussion of the
Lorentz force until §6.6. This equation without modifying forces is often

referred to as Euler's equation. To fully describe the flow, we require three further equations: an equation expressing mass conservation, an equation expressing the energy state and an equation for the gravitational field.

In physics, the moment of inertia, I, is the name given to rotational inertia: the resistance of an object to being spun. It is the total of the contributions of all mass elements multiplied by the square of their distances from a fixed point. Therefore, I decreases if an object contracts uniformly. Conveniently, the evolution of the moment of inertia of a cloud can be considered without any further equations. To do this, the fluid equation requires some manipulation. In brief, we take components in a specific radial direction, multiply by the radial distance and integrate over the entire mass. This transforms the acceleration into the acceleration of the moment of inertia:

$$\frac{1}{2}\frac{d^2 I}{dt^2} = 2(\mathcal{T} - \mathcal{T}_s) + \mathcal{M} + \mathcal{W}, \tag{4.4}$$

where for the hydrostatic configuration, we balance three quantities:
(1) twice the kinetic energy $2(\mathcal{T}$ which is in excess of the background cloud energy $\mathcal{T}_s$),
(2) the gravitational energy, $\mathcal{W}$, and
(3) the net magnetic energy, $\mathcal{M}$. This is a form of the virial theorem. The inertia term reflects changes in the size and shape of the cloud. For fast cloud evolution, this term is as large as the other terms. In the slow cloud evolution discussed here first, we ignore this term.

4.3.2   *Long-lived isolated clouds*

Now we can explore the properties of a cloud in a steady or slowly-evolving state. We can also investigate the conditions which will permit the moment of inertia to oscillate around some constant value. We would like to know what observable parameters to measure to identify a fixed cloud.

The general answer is that to maintain a cloud in equilibrium we require the pressures to balance. To show this we write all the terms in Eq. 4.4 in the form of pressures. A pressure is a force per unit area and we normally consider it to be exerted upon an area (such as the soles of our feet). Here, however, we shall also define equivalent gravitational and turbulent pressures. The point is that it takes a specific amount of energy in a unit volume of fluid to produce a pressure on the surface. In this sense, the two are equivalent. First, the thermal pressure is proportional to the number of particles and their temperature, $P_t = n_p kT$, where k is the Boltzmann constant. We take $n_p = n(H_2) + n(He) = 0.6\,n$, constituting $0.5\,n$ from the molecular hydrogen gas and $0.1\,n$ contributed from helium atoms. We also average over the entire cloud and take a symbol ˆ to denote a cloud average. The pressure is then written as

$$\hat{P}_t = \hat{\rho} v_{th}^2, \tag{4.5}$$

where v_{th} is given by Eq. 3.9. The symbol ˆ denotes a cloud average.

We can also include pressure support from small-scale swirling motions by taking a *kinetic pressure* as the sum of thermal and turbulent pressures. Together they produce our observable 1D velocity dispersion σ through the relation

$$\hat{P} = \hat{\rho} \sigma^2 \tag{4.6}$$

giving a total kinetic energy of $T = \frac{3}{2} \hat{P} V_c$ where V_c is the cloud volume.

The total gravitational energy depends on the detailed cloud structure. We usually assume it corresponds closely to that of a uniform sphere which takes the value

$$W = -\frac{3}{5} \left(\frac{GM^2}{R} \right). \tag{4.7}$$

We now write $W = -3\,P_G V_c$ to define an equivalent gravitational pressure P_G.

Collecting the terms together (ignoring for now the magnetic energy), we are left with a very simple equation:

$$\hat{P} = P_s + P_G. \tag{4.8}$$

That is, in virial equilibrium the cloud pressure stands in balance against the surface and gravitational pressures.

In this picture, a cloud is *gravitationally bound* if the gravitational pressure exceeds the surface pressure. Alternatively, a cloud is *pressure confined* if P_s exceeds P_G. Note, however, that if the cloud's turbulent motions were on a large scale, then there would be no external means of counteracting the accompanying pressure variations and so the cloud would disperse (unless the turbulence decays quicker).

We can now evaluate the terms. The surface pressure on a GMC is the interstellar pressure. In the classical picture of the ISM, there is a pressure balance between the three atomic components. In CGS units of pressure, we require about 4×10^{-12} dyne cm^{-2} to support the weight of the hot galactic corona above the Galactic plane. This, in part, consists of cosmic rays and magnetic pressure, which also pervade the GMC. This probably leaves about 2.5×10^{-12} dyne cm^{-2} as applied surface pressure applied to a GMC.

Evaluating the gravitational pressure for a spherical cloud from Eq. 4.7, we obtain a dependence on just the average column density,

$$P_G = 2.0 \times 10^{-11} \left(\frac{N}{10^{22}\,\text{cm}^{-2}} \right)^2 \text{dyne cm}^{-2}. \qquad (4.9)$$

The column is related how dark the cloud is through Eq. 2.5 and is given by Eq. 3.12. Therefore, the gravitational pressure is significantly higher than the surface pressure for the given typical extinction column. Consequently, *if* GMCs are in equilibrium, they are gravitationally bound rather than pressure confined, within this treatment as classical stable entities.

The easiest way to test a cloud for binding or confinement is to use the *virial parameter* α, 3.14, which is more or less equal to the critical pressure ratio $\hat{P}/P_G$. As discussed in §3.4, most clumps are found to be unbound but some, perhaps containing most of the mass, are potentially bound objects. The exact value for which clouds are bound, however, is more theoretical than practical, given the distorted cloud shapes and non-uniform structures within GMCs.

4.3.3 *Bound globules*

We now show that, even if pressure *equilibrium* is established, only certain clouds will be *stable*. The properties of clumps and isolated globules can be measured with less confusion surrounding their shape and size. We can use the virial theorem to predict the structure of self-gravitating spherical clumps which are isothermal. Although still rather ideal, such an analysis provides a first comparison. As it turns out, the comparison can be surprisingly good.

For a cloud to be in equilibrium we require a surface pressure

$$P_s = \frac{3M\sigma^2}{4\pi R^3} - \frac{3aGM^2}{20\pi R^4}, \qquad (4.10)$$

where we have substituted values into Eq. 4.8 from Eqs 3.14 and 4.6. Here a is a constant close to unity, inserted to account for departures from uniformity. Isothermal is here taken broadly to mean a constant velocity dispersion σ.

We deduce that kinetic pressure is dominant if R is large but that gravitational pressure is critical if R is small, for a cloud of fixed mass. We now make the following experiment, being guided by Eq. 4.10. Suppose that the external pressure, P_s, increases. The cloud can respond in two ways. If the radius is large then the term for kinetic pressure dominates. The pressure increase means the cloud shrinks. On the other hand, if the cloud is small, then the external pressure change would be met by a change in the gravitational pressure. However, that implies the opposite reaction: the radius *increases*. This is physically impossible since there has been an increase in the pressure acting across the cloud surface. To match this, the cloud density must rise and so the cloud must contract. Consequently, if we begin with a large cloud and slowly increase the external pressure, the cloud remains stable as it contracts until a particular radius is reached. A further increase in pressure then triggers the instability: the cloud collapses.

For a fixed external pressure, we obtain the maximum mass for an isothermal sphere to be stable from Eq. 4.10:

$$M_{BE} = 1.18 \frac{\sigma^4}{(G^{3/2} P_s)^{1/2}}, \qquad (4.11)$$

which is called the Bonner-Ebert mass (the appropriate value of 'a', involves solving the hydrostatic form of Eq. 4.3 and Eq. 4.14 below). With this mass, a stable sphere has its greatest central concentration. The core density is 13.98 times higher than the density just inside the surface. In terms of solar masses, this mass is quite low:

$$M_{BE} = 0.96 \left(\frac{T}{10\,\mathrm{K}} \right)^2 \left(\frac{P_s}{2\ 10^{-11}\ \mathrm{dyne\,cm}^{-2}} \right)^{-1/2} \mathrm{M}_\odot, \qquad (4.12)$$

which indicates that cold cores of high mass (e.g. $> 5\,M_\odot$) are unlikely to be supportable by thermal pressure alone. The typical high surface pressure *inside* a GMC has been taken. Corresponding to the mass is the critical radius for marginal stability,

$$R_{crit} = 0.41 \frac{G\,M_{BE}}{\sigma^2}. \qquad (4.13)$$

Clouds larger than R_{crit} are stable.

These values provide initial conditions for some theories of cloud evolution which begin from hydrostatic equilibrium. How the collapse can proceed will be discussed in §6.3.

4.3.4 *The Jeans mass*

Theoretically, stars can condense out of an almost perfectly uniform gas. Even where turbulent motions are negligible, clouds could still form. This requires an instability to be present. One instability is caused by a break-down in the delicate balance between heating and cooling which can occur under some conditions. When a gas cloud is thermally unstable, a local small perturbation will cool slightly faster than its surroundings. Its drop in pressure causes it to collapse. The fact that the density increases usu-ally causes it to cool even faster. The collapse thus accelerates and a tiny perturbation grows into a discrete cloud.

Our interest is in obtaining clouds which can gravitationally collapse into a protostar. This can be achieved directly through the Jeans instability: disturbances to regions containing sufficient mass will separate out from the uniform background.

To show this, we need to delve into a 'stability analysis'. We eliminate the gravity term **g** in Eq. 4.3 by substituting the density field ρ which produces it. The gravitational influence of all the material on each fluid particle is much more conveniently written as Poisson's equation,

$$\nabla \cdot \mathbf{g} = -4\pi G \rho. \tag{4.14}$$

If we inspect this, we notice that a uniform medium actually generates gravitational acceleration in a particular direction! This abnormality is the 'Jeans swindle', invoked by the assumed design of a truly uniform and static medium filling our entire Universe. We search for solutions to Eq. 4.3 involving small sinusoidal oscillations of frequency ω and wavelength λ. Fortunately, the cheat dissolves when we consider these first order effects. The result is a form of a *dispersion relation*:

$$\omega^2 = c_s^2 \left(\frac{2\pi}{\lambda} \right)^2 - 4\pi G \rho. \tag{4.15}$$

where c_s is the sound speed, the speed of waves in the absence of grav-ity. Ignoring the final term, we have the usual relation for a sound wave, which is non-dispersive (substituting $c_s \sim 331$ m s^{-1} and middle C of ω = 262 Hz yields $\sim$ one metre waves). Hence, this equation demonstrates the propagation of acoustic waves when modified by self-gravity. The two terms on the right represent 'restoring forces' which attempt to re-expand compressed regions. For this reason, for long wavelengths,

$$\lambda > \lambda_J = \sqrt{\frac{\pi c_s^2}{G\rho}}, \tag{4.16}$$

gravity overcomes the acoustic restoring force. This is called the Jeans criterion for gravitational instability, after the work of Sir James Jeans, published in the early 20th century. Taking a spherical disturbance this directly translates into a minimum mass

$$M_J = \frac{4\pi}{3}\left(\frac{\lambda_J}{2}\right)^3 \rho = \frac{\pi}{6}\left(\frac{\pi}{G}\right)^{3/2} c_s^3 \, \rho^{-1/2} \tag{4.17}$$

or

$$M_J = 2.43 \left(\frac{T}{10\,\mathrm{K}}\right)^2 \left(\frac{P}{2 \times 10^{-11}\,\mathrm{dyne\,cm^{-2}}}\right)^{-1/2} \mathrm{M}_\odot. \tag{4.18}$$

The Jeans mass exceeds that contained within Bonnor-Ebert spheres with the same surface conditions. This reflects the fact that a Bonnor-Ebert sphere contains matter of exclusively higher density than the surrounding medium whereas only some of the material within a Jeans length will collapse. In terms of the hydrogen number density, n, the Jeans mass is

$$M_J = M_1 \left(\frac{T}{10\,\mathrm{K}}\right)^{3/2} \left(\frac{n}{10^5\,\mathrm{cm^{-3}}}\right)^{-1/2} \tag{4.19}$$

for a molecular gas, where $M_1 = 1.18\,\mathrm{M}_\odot$ under the above assumptions but takes a value $M_1 = 0.78\,\mathrm{M}_\odot$ under a full spherical analysis. Sometimes, the Jeans mass is defined as $\rho\lambda_J^3$ and then $M_1 = 2.3$–$3.4\,\mathrm{M}_\odot$, depending on the particular assumptions.

According to a classical picture originally proposed by Fred Hoyle, a cloud or a part of a cloud with mass above the Jeans mass collapses under gravity. The Jeans mass falls as the density increases provided the collapse remains isothermal. Therefore, each part will also fragment and fragment again. A hierarchy of cloud sizes is systematically created until the collapse ceases to be isothermal. At the point when the gravitational energy, released through compressive heating, cannot escape the system in the form of radiation, the temperature of the fragments rises. According to Eq. 4.19, the Jeans mass now also rises and so the fragmentation stops. This is called 'opacity-limited fragmentation' since the heat is trapped when the core becomes opaque.

One observed inconsistency with this scenario is that the minimum Jeans mass, given the opacity due to dust, is only about 7 Jupiter masses $(0.007 \, M_\odot)$, much smaller than the observed characteristic stellar mass near $1 \, M_\odot$ (see §12.7). Furthermore, simulations show that the efficiency of the process is drastically reduced through the overall collapse of the embedding medium and non-linear damping effects as the speed of collapse becomes high. Far more probable is that bound cores develop out of the large range of transient structure in the cloud produced by strong turbulence. Nevertheless, the Jeans criterion is still a central feature in all the theories, providing the critical mass which must accumulate to initiate gravitational collapse.

4.4 Summary

In the classical theory for star formation we first searched for equilibrium between self-gravity and pressure gradients. Astrophysicists took into account a contribution from turbulence in the form of a turbulent pressure which acts in the same way as thermal pressure. That is, as a microphysical concept rather than a fluid dynamics process. An effective sound speed then replaces the true sound speed. Such configurations may serve as initial conditions from which we can simulate the collapse all the way to a star.

Do we find that clouds evolve towards static states in order to provide the proposed initial conditions? In general, approximate virial equilibrium has been confirmed in the GMCs as a whole. But, as we have emphasised, the GMCs are non-equilibrium distorted structures which evolve rapidly.

On the other hand, quiescent molecular clumps and cores do exist which appear to roughly match the static solutions discussed above. The dilemma remains of whether these could be required states or just exceptional states produced by chance in dynamic accidents.

Chapter 5

Turbulence

Asking the right question is more important than presenting an answer. We have been determined to discover the initial conditions leading stars. In the last chapter we looked at static conditions. The more we explore our Galaxy, however, the more difficult it becomes to find static initial states. This leads us to study molecular clouds as dynamic, in which initial conditions are neither prescribed or necessary. And this leads us to a new emerging paradigm which solves many of the nagging problems of the last two decades. The birth of stars is regulated by supersonic turbulence and its interplay with gravity.

Supersonic turbulence is indeed observed to dominate the dynamics of star forming environments. Molecular clouds contain random supersonic motions at an energy level capable of supporting a cloud. Turbulence is also recognised to be the key process on many scales. Theoretically, it is not easy to stop flows from becoming turbulent. It allows angular momentum to be transported at much higher rates than normally allowed by the viscous nature of the fluid. In the disk of spiral galaxies, turbulence can transport angular momentum. In the accretion disks which form around the protostars and young stars, turbulence is believed to be the agent which permits the rapid accretion (§9.4.2). What drives these motions? Are these motions maintained or do they decay?

First we must ask: what actually is supersonic turbulence? This question was avoided for 50 years possibly for what seemed a good reason: it wouldn't exist on astrophysical time scales. Any flow faster than the sound speed would be rapidly brought to a halt. The entire energy would be turned into heat and then dissipated within narrow collision zones called shock fronts. So why study something that can hardly exist? There are two good reasons. First, the interstellar medium is violent; turbulence can be regenerated as fast as it decays. Secondly, astrophysical time scales are much shorter than predicted in the absence of the fast decay. In other words, it is self-fulfilling.

A consistent theory for turbulence has defied our exertions although, amongst a superfluity of models, we have found some insight. The turbulence in space is, however, different from what we experience on flights through atmospheric clouds and new tools have been developed to understand what it does to hinder or promote the conception of stars. Having waited 50 years, supercomputers are now just capable of simulating turbulent clouds on sufficient scales and for reasonable lengths of time. The first generation of simulations have opened up new vistas.

The motions in molecular clouds are impressively fast and we find that in classical theory we cannot keep up. There is no time to achieve the required pressure balance. We need a model for dynamic activity. The motions produce ram pressure, high-pressure impacts and steep pressure gradients which feed back on the motions, generating a *turbulent system*. Within this system, some clumps and cores may develop which are close to equilibrium. These will evolve slowly, giving us greater opportunity to detect them but diverting our attention away from the vast sea of rapidly disappearing structures.

5.1   Concepts of Turbulence

Turbulence is the generic name given to a collection of random motions on a range of length scales. It can be weak, with little kinetic energy within the motions as compared to ordered or laminar motions. In molecular clouds, we observe strong turbulence: the turbulent energy dominates not only the thermal energy but also that of ordered motion such as rotation or shear on all scales with the exception of the injection scale.

Some behaviour of turbulence in star-forming regions is remarkably similar to and others wildly different from that usually encountered on earth. Most obvious is the high compressibility. The wide range in density arises from two factors. First, the cloud is a gas rather than a liquid. Second, the high speed of the gas motions result in the gas being rammed, producing a high ram pressure, which dominates the thermal pressure. We describe this property by introducing the Mach number $\mathcal{M} = v/c_s$, defined as the ratio of the speed of the flow to the speed of sound waves. Given $c_s^2 = \gamma p/\rho$ (Eq. 3.9), the Mach number squared is then proportional to the ratio of the ram and thermal pressures:

$$\mathcal{M}^2 = \frac{v^2}{c_s^2} = \frac{\rho v^2}{\gamma p}. \qquad (5.1)$$

In the case of a turbulence flow, we need to define an average speed of the random motions. We take the root mean square speed, $\sqrt{<v>}$, which we equate to the three dimensional velocity dispersion.

Interstellar turbulence can be hypersonic, supersonic, transonic or subsonic. The atomic phases and the dense molecular cores often exhibit subsonic motions. For molecular clouds, the turbulent Mach numbers we derive on inserting values from Table 3.2 are $\mathcal{M} \sim 20$ for molecular clouds, $\mathcal{M} \sim 8$ for clumps and globules and $\mathcal{M} \sim 0.7$–1.5 for dense cores. The high Mach numbers imply that the flow cannot be in pressure equilibrium.

Instead of the classical picture of eddies or vortices, in supersonic turbulence sheet-like structures develop where the supersonic flows impact. The gas is swept up into these thin dense layers after passing through 'shock fronts'. The cause of shock fronts is analogous to the reason why ocean waves break on a beach. As the depth of the water falls, the critical speed at which waves remain coherent drops. The waves breaks when the wave speed exceeds this critical speed. A shock wave occurs when the sound speed falls below the speed of the flow and we try to halt the flow. It is clear that something extraordinary must happen when sound waves become trapped, unable to move fast enough to carry information upstream. The gas is then taken by surprise and the flow energy is converted from the bulk motion into thermal energy.

Therefore, it is now evident that turbulence in molecular clouds is very different from classical turbulence. It is highly compressible with strong cooling leading to enhanced density contrasts. Pressure variations are localised. The turbulence can be described as the appearance, passage and dispersal of many shocked layers, on various scales, which may interact through collisions. Until they interact, however, the shocked layers are not aware of each other's approach: they move independently. In contrast, the motions within laboratory subsonic turbulence are dominated by large-scale interactions.

5.2   Origin of Turbulence

We don't even need to stir a fluid to cause turbulence. A perfectly ordered straight flow can quite suddenly change character. The rising column of smoke from a cigarette or chimney demonstrates the dynamics. At some critical rise speed, a laminar stream switches into a mess of wisps and twisted arcs. So we have two questions here: what supplies the energy and why does it feed into turbulence?

The energy must be injected directly into large scale motions since that is where most of the energy is found in models for turbulence as well as in molecular clouds. Almost all of this energy will then be transmitted onto smaller scales. However, some can cause larger scale motion. There are several candidates for power suppliers which we will now discuss in turn. These are:

- galactic shear,
- gravitational instability within spiral arms,
- outflows from young stars inside the clouds,
- stellar winds and radiation, and
- supernova explosions.

The Galactic disk is not in solid body rotation: the rotation speed falls with distance from the Galactic centres. The differential rotation contains sufficient energy to power the cloud turbulence but we require a mechanism to transfer this large-scale motion down to the cloud scale. Such uniform disks are, in fact, hydrodynamically stable and we need to resort to effects involving the magnetic field (i.e. magnetohydrodynamics) or gravitation in order to break the stability (see §9.4.2).

Instabilities involving self-gravitation convert the gravitational potential energy releases in a collapse into turbulence. For cloud support, this fails since it is rather like pulling oneself up by one's own bootstraps. In any case, it is found that the turbulence decays in less than a free-fall time. In spiral arms, however, the gas can be swept up by the gravitational pull. Part of this energy can be transferred into turbulence.

Can the turbulence by driven from the inside? Protostars generate high speed jets and winds which we will describe in §9.8. These outflows feed energy back into the cloud they formed from. Calculations show that the jets are energetically capable of supporting clouds on the scale of parsecs under some conditions. First, there has to be a sufficient number of outflows and it is not clear that there is. Secondly, they have to be well spread in direction and location in order to influence the entire cloud whereas most outflows are collimated, and support gas only in the vicinity of their cavities. Thirdly, they should deposit their energy into the cloud rather than penetrating outside.

The fourth requirement is that the process of energy transfer has to be accomplished with reasonable efficiency, which appears unlikely. Momentum transfer from the high speed jet into the low-speed turbulence is inefficient with most of the energy remaining in the outflow cavity or lost to radiation. The alternative would be an energy-driven flow which can, in principle, transfer a reasonable proportion of energy, around 20%. This, however, requires high-pressure hot cavities to drive the flow. Such cavities are not widespread.

Stellar winds from hot stars are powerful. Nevertheless, given a distribution in stellar masses, we can make a good direct comparison between the strength of these winds and the combined strength of the supernova explosions which will then also occur. The supernovae win out provided sufficient time has elapsed since the stars formed for the supernovae events to have taken place.

The ionising radiation from clusters of massive stars dominates the regions around them, producing H II regions and driving shocks into the surrounding neutral medium (§11.2). Energy estimates demonstrate they may be important to local support and dispersal of star-forming clouds but they are do not contribute significantly to the turbulence.

All the arguments lead us to suggest that supernova explosions must be the culprit. In addition, there is circumstantial evidence. The rate of supernova explosions in the Galaxy is estimated to be one every 50 years. The energy input is of order 10^{51} erg per event. Therefore, supernova energy is available at the level of 6×10^{41} erg s^{-1}. In comparison, the required input is estimated as follows. Given the total molecular mass in the Galaxy of $M_G = 3 \times 10^9 \, M_\odot$, a turbulent speed of $\Delta v = 6 \, \mathrm{km \, s^{-1}}$ and an average lifetime of, say, $t_C \, 4 \times 10^6$ yr, yields the dissipation rate $M_G \Delta v^2 / (2 t_C) \sim 10^{40}$ erg s^{-1}. Hence, we require just a few per cent of the supernova energy to supply the cloud turbulence in our Galaxy. The energy may be diverted vertically out of the Galactic disk, some lost to radiation, some lost to turbulence on larger scales than the GMCs and to atomic clouds. Hence direct evidence is missing. Nevertheless, it may well be that the shock waves from supernovae, ploughing through the interstellar medium, sweep up the gas into molecular clouds and thus directly control both the GMC sizes and the turbulence injection scale.

5.3   The Transfer to Turbulence

Why is the energy channelled into turbulence? The kinetic energy could now (a) remain on the largest scale, (b) dissipate into heat via viscosity or (c) be fed into turbulence. For the energy of a shear or laminar flow to be fed into turbulence, we require an instability. As with a wind blowing over the ocean, a fluid instability, called the Kelvin-Helmholtz instability, creates a centrifugal force over the ripples. This acts to amplify the wave amplitude. Without a stabilising force, which could be gravity or magnetic tension, the shear generates vortices, and these vortices will also contain shear motions with neighbouring vortices. The sheared flow breaks down into a full spectrum of vortices containing all possible sizes. Instead of using vortex size, we describe the turbulent field as a superposition of waves of wavelength λ with typical velocity, $v(\lambda)$ and energy amplitudes $E(\lambda)$.

The gas is not necessarily hydrodynamically unstable. There are other forces which damp disturbances, restoring the shear flow. In accretion disks, centrifugal forces and the gravitational force of the central protostar promote stability. There are, however, a variety of potential instabilities and it would be surprising to uncover a turbulent-free flow. Besides thermal and gravitational instabilities already discussed, there are viscous and magneto-rotational instabilities (see §9.4.2). Moreover, the thin compressed shells of swept-up gas behind supernova shock waves are prone to warping, distortion and fragmentation.

The eddy motions will certainly not be produced if viscosity smears out the relative motion. If the mean free path of individual gas particles, λ_p, exceeds the length scale of the shear motion, L, then the energy will be directly diverted through particle collisions into heat. The viscous force was neglected from the inviscid equation of motion, Eq. 4.3. We now extend the equation to include it. Omitting other forces, we then have the Navier-Stokes equation:

$$\rho\frac{d\mathbf{v}}{dt} = -\nabla P_t + +\rho\nu_m\mathbf{\Delta}v \qquad (5.2)$$

where the kinematic viscosity is, as we would expect, proportional to the mean free path and the thermal speed, (see Eq. 3.9):

$$\nu_m = \lambda_p v_{th} \sim 2 \times 10^{17} \left(\frac{n}{100 \, \text{cm}^{-3}} \right)^{-1} \left(\frac{T}{10 \, \text{K}} \right)^{1/2} \text{cm}^2 \, \text{s}^{-1} \qquad (5.3)$$

where the evaluation has used Eqs. 2.3 and 3.9.

The symbol Δv represents a *diffusion* operator. Diffusion means that the force on a fluid element is not proportional to the gradient in velocity, but the change in this gradient with distance. That is, there would be no viscosity if the shear is linear because the effects of higher speed on one side and lower speed on the opposite side of the fluid element would cancel each other. (Similarly, thermal conduction acts through diffusion: if a warm location is heated on one side and cooled on the other, then it can maintains its temperature.) On average, the collisions would not alter the momentum. Hence, the viscous force depends on gradients of gradients.

The implication of the diffusive nature is seen from the dimensions. When we have velocity changes of order U over a length scale L, we expect velocity gradients of order U/L and a dynamical time of order L/U. Advection or acceleration is of order U^2/L and viscous acceleration is of order $\nu_m \, U/L^2$. Hence, viscosity is most effective on small scales.

This is traditionally expressed by the Reynolds number. Viscosity will damp fluid instabilities if the viscous force is sufficiently large relative to the non-linear advection term. The ratio of these terms in Eq. 5.2 is measured by the Reynolds number

$$\mathcal{R}e = \frac{U \, L}{\nu_m}. \qquad (5.4)$$

The Reynolds number, being dimensionless, is relevant to flows on any scale: we can scale up laboratory experiments to predict Galactic-scale behaviour. So to decide if clouds are viscous, we only have to evaluate $\mathcal{R}e$. We find $\mathcal{R}e \sim 10^9$ for the giant molecular clouds and $\mathcal{R}e \sim 10^8$ for small dense cores. In the laboratory, viscosity can maintain a laminar flow even for $\mathcal{R}e$ of order several thousand. In clouds, however, turbulence appears unstoppable.

Note however, that we have assumed that the cloud is being sheared by external forces over a wide extent. A jet flow from a protostar, even when as extended as the cloud, may maintain a laminar flow in itself and the cloud. This is because turbulence does not readily spread: it is confined to thin boundary layers. The process that occurs here is of utmost importance to star formation: the role of molecular viscosity is replaced by a turbulence viscosity. At high $\mathcal{R}e$, turbulence becomes much more efficient at transporting the momentum between locations. The eddies effectively decrease the Reynolds number to unity within small volumes. Nature thus substitutes turbulent viscosity for molecular viscosity and the zone of turbulence is predominantly confined to narrow shear layers. Hence, disturbing the surface of a uniform cloud would not be sufficient for the turbulent motions to be transferred internally.

We conclude that we have the power, the time and the means for molecular clouds to be filled with supersonic turbulence during their formation. After formation, however, it is not easy to resupply an entire cloud.

5.4   Dissipation of Turbulence

In this section, we determine the fundamental properties of turbulence which ultimately control the nature of star formation. Yet we cannot predict the fate of a particular cloud. The instantaneous behaviour of a turbulent flow is unpredictable. So we resort to averages and a statistical description. First, we examine the energy budget. In addition, we search for correlations in the velocity field on each scale as begun in §3.4. Obviously, we expect the velocity difference, Δv, of distant points to be uncorrelated and close points to be highly correlated. In more detail, we correlate many powers of the velocity difference to try to characterise turbulence. These are called *structure functions*.

The theory of classical turbulence remains relevant. We take here an heuristic approach to capture the basic behaviour. First, we stir the fluid: we inject the energy on a relatively large scale. This generates large scale eddies, the vortex motions which can be quite long-lived. The cortices interact, producing smaller and smaller vortices until at some very small scale, a vortex is smeared over by viscosity. At this scale, the mean free path of molecules or atoms breaks down the fluid approximation and particles collide. The collisions convert the energy in the eddies into thermal energy i.e. heat.

A steady state is attained in which the power injected at some wavelength λ_{in} is redistributed into a spectrum of eddies or waves with speeds v which depend on the wavelength λ. The wave spectrum continues down until a dissipation scale λ_{out} is reached. The time scale for transfer should only depend on these parameters and so must be proportional to $t_e = \lambda/v$. Therefore, the specific energy $E_e = v^2/2$ is transferred between scales at the rate $\eta v/\lambda$ where we expect η to be of order unity. This implies that the rate at which vorticity redistributes the energy on each scale is determined by an *eddy or turbulent viscosity* which is

$$\nu_e = \lambda \times v. \tag{5.5}$$

In this manner, the energy injected is transferred through the vortices according to a law

$$\dot{E} = \frac{\eta E_e}{T_e} = \frac{\eta v^3}{\lambda}. \tag{5.6}$$

Therefore, in a steady state, the vortex motions satisfy $v \propto \lambda^{1/3}$. This yields the classical energy spectrum of incompressible turbulence, first worked out by Kolmogorov in 1942: $\mathcal{E} = v^2/2 \propto \lambda^{2/3}$.

This may prove to be the most critical formula in determining the properties of our Universe. It implies that most energy is contained in the largest scale eddies. This also implies that turbulence can support a cloud on it's largest scale provided that we add sufficient energy on the cloud scale and resupply this energy on a time scale of order L/U. Adding the required energy on smaller or larger scales does not provide the necessary support. Furthermore, insertion of a turbulent 'micropressure' which behaves as an amplified thermal pressure is now seen to be invalid. Any attempt to add turbulence to the fluid equations as an approximation must take into account the dependence on wavelength.

We can now determine the minimum size of the eddies. The Reynolds number based on the input scale is $\mathcal{R}e_c = U\lambda_{in}/\nu_m$, assumed to be extremely large. The effective Reynolds number now depends on the wavelength and will be proportional to $v \times \lambda \propto \lambda^{4/3}$. Therefore, there will exist what is termed an 'inertial range' in which the energy is transferred by fluid motions from the input scale λ_{in} down to the scale where the effective Reynolds number is unity:

$$\lambda_{out} = \lambda_{in}/\mathcal{R}e_c^{3/4}. \tag{5.7}$$

Therefore, for the Reynolds numbers quoted above, we can expect inertial ranges extending six orders in magnitude. Once vortices approach the minimum size, viscosity is significant and the energy in the bulk flow is transferred into heat (before being radiated away in the interstellar medium).

The maximum size of eddies which contain a non-negligible amount of energy is close to the input size. Some energy is also transferred or 'backscattered' to longer wavelengths but very inefficiently. If the problem allows larger vortices to develop, they certainly will but they will also break down swiftly. Calculations show that if there is no energy initially present, a spectrum of the form $\mathcal{E} = \propto \lambda^{-5}$ will develop. Thus, very little energy can be channelled into the so-called 'infrared' regime. This is what makes it so extremely difficult to hinder cloud collapse on large scales by the addition of fast-driven small-scale turbulence (through jets, for example).

Supersonic turbulence is in many respects different from subsonic turbulence. Motions on the largest scale can be decelerated abruptly and the energy directly turned into heat in narrow shock waves. There is not the same need for an energy cascade through an inertial range. Nevertheless, most of the energy on any scale can be found in solenoidal rather than compressional waves. It would appear that the energy can remain a few times longer in vortex-type motion before either breaking down into smaller vortices or being directly transferred into compressional modes and dissipated.

Supersonic turbulence is *spatially local*: no warning is communicated to material approaching the shock front of what is about to happen. In contrast, subsonic turbulence is *spectrally local*, referring to the wavelength and the fact that an eddy is much more likely to interact with eddies of similar wavelength. For these reasons, our understanding of supersonic turbulence has relied heavily on computer simulations.

5.5 Computer Simulations

How does supersonic turbulence in molecular clouds differ from subsonic turbulence? To find the many answers, we investigate through computer simulations. We commonly apply two rudimentary methods to solve the hydrodynamic equations under cloud conditions. Both methods have their limitations, some of which become less disturbing when we can employ supercomputers with massive memory and prodigious speed.

Grid methods substitute a three-dimensional grid of fixed points for a cloud. In this Eulerian approach, the book-keeping is done by accounting for the changes in quantities such as the mass, momentum and energy over a huge number of tiny time intervals. This involves calculating fluxes through the cell surfaces defined by the grid and adding on sources and sinks. This method has great physical versatility, allowing extensions to magnetohydrodynamics and ambipolar diffusion (§6.6). It can also reliably simulate shock wave structures and provides efficient resolution of tenuous regions. Elegant schemes have been invented to aid the tracking and resolution of collapsing regions. One method employs multiple grids: a mesh within a mesh. Another method, called Adaptive Mesh Refinement, only inserts a finer grid where needed and removes finer grids when obsolete.

A second method represents the fluid as a huge swarm of gravitating particles and keeps track of their locations and fluid-like interactions. This Lagrangian approach, called Smoothed Particle Hydrodynamics (SPH), permits a wide range in density and size scale to be reached with high precision. SPH concentrates its computing effort on the high density regions.

SPH also has the advantage that the constituent SPH particles can be replaced with a single 'sink particle' wherever we are sure that a compact bound object has formed, allowing the procedure to continue. These sink particles are substitutes for newly formed stars, interacting thereafter only through gravitational forces and by accretion of gas particles that fall into their sink-radii. In SPH, the fate of a sink particle is cast whereas in the grid methods, the contents of a single zone can be easily dispersed.

In fundamental studies, energy is injected randomly throughout a region at an initial time and then simply allowed to decay. The velocity distribution begins by steepening up as wave 'crests' of high pressure move faster and catch up with the low-pressure 'valleys'. The overtaking is a result of the higher sound speed in the wave crests. The steepening continues until a sharp velocity difference ΔU over a width ΔL develops such that the parameter $\mathcal{R}e = \Delta L \Delta U / \nu_m \sim 1$. In other words, the wave steepens until viscosity is capable of dissipating the wave energy. This then defines the configuration for a 'collisional shock'.

A neat property of a shock front is that what comes out is fixed by what goes in. The subsonic post-shock flow is fixed by the supersonic pre-shock flow. Physically, the wave steepens until whatever frictional-type force or wave we impose reaches the correct magnitude to resist the steepening. The viscous process only regulates the width of the shock front. The smooth turbulent waves rapidly progresses to a field of shock waves. The shocks interact and fragment, generating numerous weak shocks. The example of a field of shock waves displayed in Fig. 5.1 shows that they tend to occur in pairs which temporarily sandwich dense slices of gas.

The compression in a shock depends on the Mach number. We can demonstrate this for a shock in which the gas cools back down to the same temperature it had before the shock. We suppose that the density is ρ_1 upstream of the wave moving with speed v_1 and Mach number $\mathcal{M} = v_1/c_s$. We also suppose that the shocked and compressed gas move downstream with mass density ρ_2 and speed v_2 relative to the shock front. In a steady

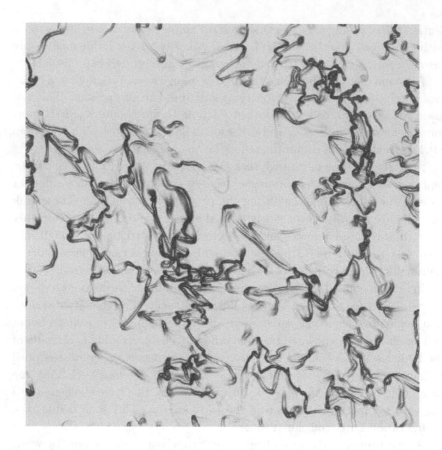

Fig. 5.1   A field of shock waves produced in a computer simulation of supersonic turbulence, in which the original turbulent energy is allowed to decay. The simulation was executed with the ZEUS-3D grid-based code.

state shock front, the mass and momentum flows are conserved. That is, through unit area of the shock front,

$$\rho_1 v_1 = \rho_2 v_2 \tag{5.8}$$

and

$$\rho_1 v_1^2 + p_1 = \rho_2 v_2^2 + p_2. \tag{5.9}$$

Note that the pressure is increased across the shock front from $p_1 = \rho_1 c_s^2$ to $p_2 = \rho_2 c_s^2$. Energy is not conserved since we have assumed that the gas heated in the shock is immediately cooled (lost by radiating). The isothermal condition means that the sound speed is fixed.

Classical formulae for isothermal shocks are obtained on manipulating these equations: $v_1 v_2 = c_s^2$ and a shock compression

$$\rho_2/\rho_1 = \mathcal{M}^2. \tag{5.10}$$

(A second solution always present is $v_2 = v_1$ i.e. the absence of any event.) Therefore, hydrodynamic turbulence with a quite moderate mean Mach number produces sheets of very dense gas.

Subsequently, as the material external to the shocked layers is gradually swept up, only lower speed gas remains to be consumed. Hence, ΔL increases and the layers of shocked material expand. The shocked layers are sheet-like. The sheets warp, buckle and fragment as well as collide with each other. Hence, a cascade of sorts still develops. The velocity difference as a function of distance λ now takes the statistical form close to $\Delta v \propto \lambda^{1/2}$, consistent with a random distribution of both positive and negative jumps in velocity.

In decaying turbulence, we express the decay as a power law in time with a decay time L/U and a power law index α ($\mathcal{E} \propto t^{-\alpha}$) which depends on the particular conditions. The decay is effected by an increasing number of very weak shocks while all the strong shocks disappear. Most numerical simulations yield $\alpha \sim 0.8$–1.3. The value depends on the set up of the boundaries, the physics and the energy injected. A power law description may not always be valid.

A second possibility is that the turbulence is regularly driven from one of the sources discussed above. The driving regulates the average speed and Mach number of the turbulence, enabling the energy to be dissipated at the rate it is introduced. The energy dissipation occurs through a small number of large-scale strong shocks. But can the driving support a global collapse?

5.6 Collapse of Turbulent Clouds

We can derive a turbulent dissipation time to compare to the gravitational free-fall time of Eq. 4.1. In all cases, turbulence does not resist gravity for long: the decay time is even thought to be shorter than the free-fall time. The magnetic field will reduce the effective Mach number we should employ here and, although uncertain, may well make the two timescales comparable.

Therefore, the supersonic turbulence that we observe must be continuously driven or clouds cannot be older than a free-fall time. As discussed in §4.1, clouds are typically only a few free-fall times old, suggesting a modicum of driving. It should also be noted that the collapse itself cannot muster enough turbulence to provide significant self-support even though the motions are amplified as potential energy is released.

So how does a complex network of interacting shocks form stars? Most gas accumulates in layers only to be dispersed again. However, particularly mass concentrations will be compact enough to be gravitationally bound. These will collapse under gravity. If the mass exceeds about the Jeans mass, given by Eq. 4.17, collapse begins. That is not all, however. For collapse, we require (a) the Jeans mass to be exceeded within the shock layer, (b) the fragmentation to be sufficiently rapid so that the compressed layer is still maintained by the ram pressure of the instigating shock and (c) the accumulation and the collapse time to be less than the time interval between the arrival of successive shocks. A region may then gravitationally decouple from its surroundings. Further gas may then accrete on to this region to build up the available mass for the construction of the star.

Continuously driven turbulence is thus more likely to prevent star formation while intermittent turbulence promotes it. The driving tends to provide global support with some inefficient star formation taking place in isolation. We find local promotion but global prevention. This scenario has been used to interpret isolated and low-mass star formation regions such as in Taurus.

To completely prevent star formation, sufficient turbulent energy must be introduced to support the small scale structure. As we have seen, this is not easily achieved through an energy cascade since the energy on the small scales dissipates rapidly. We would thus need a lot of energy to waste. Furthermore, we have no other mechanism to provide small scale turbulence except in locations pervaded by outflows from already-formed protostars. These outflows could indeed regulate star formation on these scales.

So how does the cloud evolve? As gravity begins to take over, the general features of the structure of supersonic turbulence are as follows:

- The layers usually tumble. They possess angular momentum about an axis in the plane of the layer.
- The layers fragment due to motions within the plane caused by self-gravity. A network of striations with mean separation given by the local Jeans length ensues.

- A striation evolves into a string of cores with a typical separation also given by the Jeans length.

The isolated cores which form move quite fast, with the mean speeds of the accumulated turbulent material. Less mass is left in the diffuse cloud and the cores begin to interact gravitationally. Close encounters between cores become critical.

Decaying turbulence appears to be the means to get an entire cloud to collapse: to form a cluster containing high-mass stars. We need weak turbulence on the appropriate scales so that the Jeans mass is far exceeded and collapse followed by fragmentation can occur. With no birth control, this should occur rapidly – in a free-fall time. In this manner an entire cloud could begin forming stars.

A simulation which combines decaying turbulence and gravitational interactions is displayed in Fig. 5.2. Fragmentation is allowed down to the expected scale of the opacity limit, (see §4.3.4). The cloud contains $50\,M_\odot$ within a diameter of 0.375 pc. The free-fall time of the cloud is 190,000 years. Note the chaotic gas flows and dynamical stellar encounters. Many of the stars and brown dwarfs are surrounded by disks which are truncated during encounters. Mass supplies can be cut off and stars ejected through close encounters at any time. Such competitive environments are discussed in §11.9.

How is star formation within a cloud terminated? There are several possibilities and we believe we observe all of them in action. First, an external flow can penetrate through the diffuse remnant cloud gas, sweeping it clean. If the cloud is located immediately adjacent to a developing OB association this would seem inevitable. On the other hand, a cloud which is further away can be compressed and shaken by the winds from the OB stars, providing the turbulence to re-invigorate or *trigger* a new episode of star birth. We do indeed have strong evidence for *sequential star formation* on these scales (e.g. §3.3.4).

A second mechanism is the feedback from young stars themselves. Both their outflows and ionising radiation are sources of small-scale turbulent support. The outflows are directed, focusing their impact over limited portions of the cloud (§9.8). This concentrated effort may simply blast out material in their paths. Again, however, clumps which contain less than a Jeans mass may be pushed over the limit by a shock emanating from a nearby recently-born protostar. We thus have another potential triggering mechanism.

If all else fails, the cloud becomes exhausted. The stars which form cream off the high density locations. Most of the gas is then left in diffuse form until the molecules become dissociated by penetrating UV radiation. This gas then returns to the general interstellar medium until it gets caught again by a lurking shock wave.

5.7   Molecules in Turbulent Clouds

When we probe deeper into the properties of molecular turbulence, will we find irreconcilable inconsistencies with the observations? At present, there are some encouraging consistencies. In theory, we have shown that turbulence cascades according to a law $v \propto \lambda^{1/3}$ if subsonic but according to $v \propto \lambda^{1/2}$ if supersonic. The observed law for clouds, Eq. 3.10, corresponds closer to the supersonic law. The two extremes in scale, however, of interstellar and subsonic cores might be more consistent with the subsonic law.

There is no time for the molecules to form if molecular clouds are transient and uniform! It is reasonable to suppose that molecular clouds form from atomic clouds. Yet, we must not only compress the atomic clouds but we must also wait for the hydrogen atoms to pair up. The average time taken for molecules to form on grains and to be re-injected back into the gas phase from a neutral atomic cloud is 2×10^7 yr, on substituting the values from Table 3.1 into the formula in §2.4.3. This is plenty of time in the classical quasi-static models but a factor $4-10$ too long in the dynamical model.

To solve this problem, we first note that interstellar turbulence is different from laboratory studies of gas turbulence because of the speed at which the gas cools. The gas loses the heat energy through radiation. The density rises to compensate in order to resist the surrounding pressure. Therefore, very strong cooling in the regions of high pressure and density lead to enhanced density contrasts, producing a medium conducive to collapse. Furthermore, the density contrasts amplify the rate at which molecules can condense out of atomic clouds. As an extreme example, if we placed all the mass of a uniform atomic cloud into half of the volume, the time scale for molecule formation is halved since the reaction rate at which an atom binds with a partner is doubled. Therefore, a moderate Mach number of $4-10$ suffices to create molecules as fast as molecular clouds.

5.8 Summary: The New Paradigm

In our discussion of cloud formation we attempted to isolate physical processes and develop them as competing theories. Each may have its own realm of importance, although probably in combination with others. Comparison of diverse regions, however, demonstrates ubiquitous behaviour: self-similar or fractal structure and power-law distributions and correlations are remarkably similar.

There is one means to link these observations: by taking a multi-physics approach in which the macroscopic properties of fluid dynamics unites the system properties. *The nature of supersonic turbulence is defining or supervising the formation, structure and development of molecular clouds.*

Finally, having made progress, we step back to ask: do clouds actually exist? The classical idea was that the observed Gaussian velocity distributions correspond to the superimposed spectral lines of many small discrete components within a cloud. The deviations from the classical Gaussian profile is attributed to effects on the periphery of individual clumps.

In the turbulent picture, discrete objects don't exist. Layers of gas are constantly dispersing and new layers forming as a sea of weak shock waves pervades the entire region. The high-speed deviations are then very naturally interpreted as due to *intermittency*. Intermittency is the name given to the spatial and temporal variations that accompany turbulence due to the fact that most of the energy is contained on the largest scales: any variations to this structure are not veiled by the random contributions of many other regions. The clouds we detect using any tracer molecule are just the tips of the iceberg viewed in the light of that molecule.

It is often stated that whatever the initial conditions were, they will be washed over and made untraceable if we let turbulence evolve. Analogous to the macroscopic fluid approximation to the microscopic particle motions, turbulence can be described by a statistical theory in which initial conditions cease to have any effect. We cannot predict the final locations of the young stars although we can determine a probability.

Quiescent clouds in dynamical equilibrium may prove quite common because they live quiet stress-free lives. Like a family seen from the outside, all get on peacefully, working in unison to raise the next generation of young stars. It turns out that the active family is extremely violent: true star-forming clouds don't wait to be observed and, as we will see, the protostars which form do their best to quickly cover up their origins.

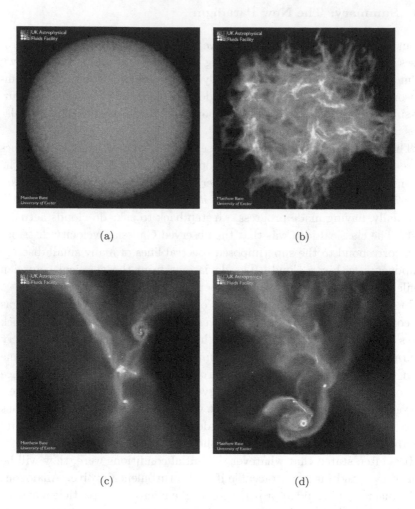

(a)   (b)

(c)   (d)

Fig. 5.2 Star formation from a massive clump. We begin with a dense core of gas
containing over $50\,M_J$ (top left) and a mean Mach number of 6.4. When sufficient
energy is lost in some regions, gravity can pull the gas together to form dense 'cores' (top
right). Stars and brown dwarfs form and interact with each other, many being ejected
from the clump (lower left). Protostars also form within the disks of circumstellar gas.
The orbits of these objects are unstable and they are quickly ejected from the binary.
These are the losers in the competitive accretion. After a pause, remaining gas gathers
again and star formation repeats. Image (d) shows the cloud and star cluster at the end
of the simulation which covered 266,000 years. These images were extracted from an
SPH simulation executed on UKAFF by M. Bate, I. Bonnell and V. Bromm employing
500,000 particles.

Chapter 6

The Collapse

A nest has been constructed and methods leading to conception have been tried out. The next event is that, from out of a turbulent cloud, the eggs are somehow laid: dense and bound cores form in regions where the turbulence has partly abated.

In this chapter, we deal with the ensuing pregnancy. In the classical problem, we wish to determine what core 'initial conditions' lead to young stars i.e. the properties which determine the fertility of the embedded eggs. In this approach, we aim to extract the physical laws which control the masses of stars. These laws are fundamental to the nature of our Universe and our existence. We now examine and analyse cores without anticipating the outcome. The constraints we could impose from the known adult star populations – the statistics of main-sequence star populations – will only be applied when the story is more developed (in §11.9).

This chapter then brings us to the final preparations: the internal adjustments before the moment of birth. The inexorable collapse may then begin.

6.1 Observing Starless Cores and Pre-stellar Cores

Starless cores are dense cores in which no protostars or stars of any kind are embedded. Yet, many cores were found to already contain embedded objects when in 1983 the Infrared Astronomy Satellite surveyed the sky in the far-infrared spectral range 12–$100\,\mu$m. The time scales and statistics provide evidence that both types of cores are transient with lifetimes of order of 1 Myr.

We define the subset of cores which are fertile as pre-stellar cores. These are the eggs in which a protostar is destined to be nurtured. On the other hand, the acronym EGG stands for Evaporating Gaseous Globule. This defines another subset of cores in which nearby massive stars are having a strong influence on egg formation and development. The influence is effected by UV heating which evaporates the outer layers and pressurises the inner layers.

Starless cores are larger in size and tend to be less centrally condensed than those already with embedded sources. They also typically contain sufficient mass to form a protostar and it's envelope. Hence they are clearly *the* potential birth sites. However, although they could be on the verge of gravitational collapse, they may equally be unbound structures about to disperse. Hence, a starless core is also a pre-stellar core if the collapse can no longer be aborted and star birth is inevitable. We can probably pick out the pre-stellar cores from them as those displaying indications of infall and/or exhibiting a central density peak.

We have only recently had the technology to describe starless cores and, obviously, the first issue we want to settle is whether they are collapsing or not. Two techniques are used: isotopic molecular line emission and dust continuum emission. Certain molecular lines such as from NH_3, CS and HCO^+, provide some evidence for inward gas motions through Doppler spectroscopy. The spectral signatures of mass infall suggest that the proto-stellar collapse has already started in some cores but these signatures only become obvious after the birth (§7.8).

The cores are very cold and so emit most of their light at submillimetre and millimetre wavelengths. At low temperatures, the continuum radiation is generated by dust particles. The dust emission is generally optically thin at these wavelengths which has the advantage that the mission is directly proportional to the total mass involved, although several factors (as always) such as the relative amounts of dust and gas (the dust-to-gas ratio) and non-uniformity may well complicate the calculation. A single dust grain emits and absorbs radiation as a blackbody proportional to an effective surface area and the well-known Planck function

$$B(\nu) = \frac{2\,h\,\nu^3}{c^2} \frac{1}{\exp(\frac{h\nu}{kT_d} - 1)} \tag{6.1}$$

where T_d is the dust temperature, h is the Planck constant and c is the speed of light. Note that the flux density B has CGS units of $\mathrm{erg\,s^{-1}\,cm^{-2}\,Hz^{-1}}$ but is usually expressed in Janskys where $1\,\mathrm{Jy} = 10^{-26}$ of the CGS value. Even so, the fluxes we detect tend to be some small fraction of a Jansky.

The blackbody curve possesses a maximum at

$$\nu_{max} = 58.8T_d \text{ GHz}, \quad \text{or} \quad \lambda_{max} = \frac{0.510}{T_d} \text{ cm}, \tag{6.2}$$

which only depends on the temperature (Wien's Law). Thus, the continuum emission from a cold molecular cloud of temperature 10 K peaks at 600 GHz or 0.5 mm, in the submillimetre regime.

Other expressions are found in the literature. For example, astronomers often prefer to plot $\nu \cdot S(\nu)$ instead of the observed flux $S(\nu)$ since this is an energy flux, and so demonstrates the spectral region containing most of the power emitted rather than where most of the photons are detected. An analogous spectral energy in terms of wavelength is often employed, $\lambda \cdot S(\lambda)$, where $S(\lambda)$ is the flux per unit wavelength.

The predicted maxima of these quantities are

$$\nu_{max} = 81.7 \, T_d \text{ GHz} \quad or \quad \lambda_{max} = \frac{0.367}{T_d} \text{ cm} \tag{6.3}$$

for blackbodies. The majority of astronomers express the *spectral energy distribution* (SED) in terms of wavelength. However, if the flux density is in Janskys, then the first expression will still apply.

Integration over the entire spectrum of a blackbody yields a luminosity which depends only on the temperature,

$$I_{bb} = \frac{\sigma_{bb}}{\pi} T_d^4, \tag{6.4}$$

where $\sigma_{bb} = 5.670 \times 10^{-5} \text{ erg s}^{-1} \text{ cm}^{-2} \text{ K}^{-4}$ is the Stefan-Boltzmann constant.

Unfortunately, trying to fit a pristine blackbody curve to yield the temperature, luminosity, and hence the mass, often fails. Observations of cores now extend over several wavebands, in some of which the core is optically thick. The radiation is then modelled with a modified blackbody spectrum. This modified spectrum is often called a greybody curve and is calculated according to the rule

$$I(\nu) = B(\nu)(1 - e^{-\tau_\nu}), \tag{6.5}$$

where the optical depth through the intervening dust τ is quite a complex function of frequency.

We often assume a power-law representation, $\tau_\nu = (\nu/\nu_0)^{\beta_o}$, for the optical thickness at frequency ν with ν_0 being selected such that $\tau_\nu = 1.0$ at $\nu = \nu_0$. The optical depth will be proportional to the column of gas, $N(H)$, and the dust absorption cross section, σ_λ. The wavelength dependence is quite complex, with one useful approximation being given by

$$\sigma_\lambda = b\frac{\tau}{N(H)} = 7 \times 10^{-25}\left(\frac{\lambda}{100\mu m}\right)^{\beta_o} \text{cm}^2, \tag{6.6}$$

where $b = 1$ and $\beta_o = -1.5$ for $40\mu m < \lambda < 100\mu m$ and $\beta_o = -2$ for $\lambda > 100\mu m$.

The dust opacity κ is commonly used in the optically thin limit $I(\nu) = \tau_\nu B(\nu)$ where $\kappa = \sigma_\lambda/m_H$ (note that dust opacity and emissivity are equivalent according to Kirchoff's law). The opacity increases as the gas particles stick onto the dust grains. For example, $\kappa \sim 0.005\,\text{cm}^2\,\text{gm}^{-1}$ at 1.3 mm in clouds, but it is typically a factor of 2 larger in dense locations such as cores and probably 4 times larger in the disks which may go on to form planets (protoplanetary disks, see §9.4.2).

In practice, β_o is available as a parameter to be determined by a fitting procedure. It embraces information about the dust type and evolution. For example, a very steep function with $\beta_o = 2.8$ was found for a massive giant molecular cloud core GCM 0.25+0.11 located near the Galactic centre. The high exponent is consistent with the presence of dust grains covered with thick mantles of ice. For the starless core L 1544, Figure 6.1 shows again that β_o is quite high in comparison to a protostellar core. For the same reason, molecular lines are unreliable as tracers of gas density in dense cores since molecules may freeze out in large quantities, forming ice mantles around refractory grain particles.

In the light of these complications, it makes sense to characterise a core by a bolometric luminosity and temperature. The bolometric luminosity is simply the total luminosity summed over the entire spectrum and the bolometric temperature is defined as the temperature of a blackbody whose spectrum has the same mean frequency $<\mu>$ as the observed spectrum. We find

$$T_{bol} = 1.25\frac{<\nu>}{100 \text{ GHz}} \text{ K}. \tag{6.7}$$

Therefore, as a core gets hotter, the mean frequency shifts upwards.

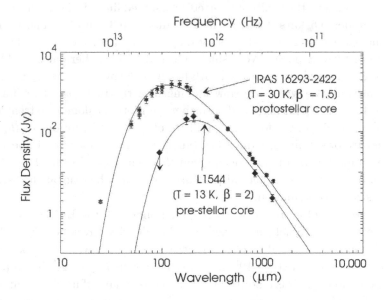

Fig. 6.1   The spectrum of the starless core L 1544 and the protostellar core IRAS 16293-2422, fitted with greybody curves. (Credit: from data presented by P. André, D. Ward-Thompson & M. Barsony in Protostars and Planets IV, edited by V. Mannings, A. P. Boss & S. S. Russell, 2000 (University of Arizona Press).)

6.2   Properties of Starless Cores

6.2.1   *Physical parameters*

Most observed cores contain sufficient mass to form stars. Derived masses range from $0.05\,M_\odot$ to $30\,M_\odot$. The smallest detected cores might go on to form brown dwarfs rather than stars since the critical mass required in order to be able to commence hydrogen burning (fusion) is $\sim 0.075\,M_\odot$ (§12.1).

Temperatures are typically under 15 K. This means that the dust is cold, consistent with the absence of a warming embedded protostar. For example, a temperature of 13 K is determined for the L 1544 starless core in Taurus. A model fit is shown in Fig. 6.1. A confident result requires data from a wide spectral region – data in this figure were acquired from several telescopes: the Infrared Space Observatory ISO, James Clerk Maxwell Telescope (JCMT, Hawaii) and the Institut de Radioastronomie Millimetrique

(IRAM, Spain), thus combining innovative technology from many nations.

Can the cores be thermally supported? Observed line widths are narrow, typically under 0.5 km s^{-1} with a mean value of 0.3 km s^{-1}. The mean value implies one-dimensional velocity dispersions of 0.13 km s^{-1}. Heavy molecules, however, possess very small line widths in a thermal gas. The lines being measured are associated with heavy molecules such as N_2H^+ (29 atomic mass units), which are slow moving in a thermal gas. Therefore, the expected velocity dispersion due to thermal motions alone is extremely small, about 0.05 km s^{-1} for a temperature of 10 K. The motions of *these molecules* are predominantly due to turbulence. The motions are, however, not supersonic but mildly subsonic, given the sound speed of 0.19 km s^{-1} from Eq. 3.9. Therefore, thermal pressure dominates but turbulence still makes a sizable contribution.

The structure of a pre-stellar core is exemplified by the L 1544 core. It consists of a dense kernel surrounded by a low density envelope (Fig. 6.2). From dust emission and absorption observations, we know that the kernel has a central density of about 10^6 cm^{-3} inside a radius $R_{flat} = 2,500$ AU (an elongated region). This is followed by a $1/R^2$ density fall-off until a radius of about 10,000 AU. This 'central flattening' is quite common, with a break to a steeper fall-off occurring at about 6,000 AU (0.03 pc). Typically, cores are not round or elliptical but roughly elongated with aspect ratios in the range 2–3. They may well be prolate or triaxial rather than oblate. Some even appear completely irregular. Hence, to describe the density distribution, a core centre is first defined as the centroid of the dust emission and the radial density distribution is then an angle-averaged quantity.

6.2.2 *Dynamical parameters*

Inward motions are inferred by observing the *infall asymmetry* in spectral lines. A necessary signature is a combination of a double peak with a brighter blue component or a skewed single blue peak in an optically thick emission line. We need to discard the possibility that two unrelated clouds are superimposed and their combined signature imitates a collapse signature. This can be achieved by checking for a simple Gaussian single peak in an optically thin line. In addition, other systematic motion – rotation and outflow – must be distinguished.

As we gather data, a picture of how starless cores evolve is emerging. Cores occasionally show evidence for infall with very low speeds in the range of 0.05–0.09 km s^{-1}. L1544 can be regarded as an archetypal pre-stellar

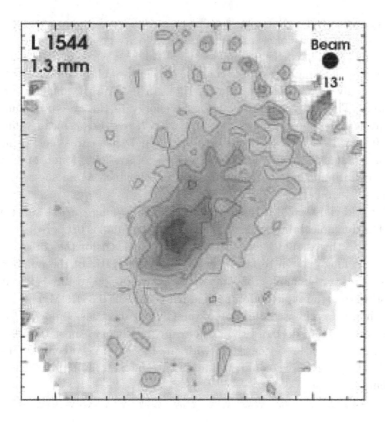

Fig. 6.2 The starless core L 1544 at 1.3 mm. (Adapted from paper by D. Ward-Thompson, F. Motte & P. André, MNRAS 305, 143 150.)

core: a starless core which is entering the stage of rapid infall. We find that the infall typically extends to radii of 0.06–0.14 pc. This extended infall phenomenon may be a necessary stage in core evolution.

The chemical composition is systematically differentiated. Near the core centres, some molecules, including CO and CS, almost vanish. In contrast, NH_3 increases toward the core centre while the N_2H^+ molecule maintains a constant abundance. These observations are consistent with a model in which CO is able to condense onto dust grains at densities above 10^5 cm^{-3}. The corresponding radius of the inner depleted region is typically $R_d \sim$ 6,500 AU.

The angular momentum of cores can be evaluated through molecular tracers such as NH_3. Cores in the size range 0.06–0.6 pc are observed to rotate in the majority of cases studied with velocity gradients in the range 0.3–3 km s^{-1} pc^{-1} (corresponding to 10^{-14}–10^{-13} s^{-1}). Here, the angular velocity Ω scales roughly as $R^{-0.4}$ and the specific angular momentum J/M scales roughly as $R^{1.6}$, with a value of 10^{21} cm^2 s^{-1} on the smallest scales measured. The ratio of rotational kinetic energy to the absolute value of the gravitational energy shows no trend with R and has a mean value of about 0.03 with a large scatter. It is also found that cores tend to have gradients that are not in the same direction as gradients found on larger scales in the immediate surroundings, an effect which again suggests the presence of turbulence.

We could constrain our theories if we knew how long starless cores survive. The dynamical time scale of those exhibiting infall is 0.1 pc/0.1 km s^{-1} $\sim 10^6$ yr. So this should be a strict minimum (at least for the cores which are present long enough to favour detection). The ratio of the number of starless cores to the number of cores with very young protostars is about 3, suggesting that the typical lifetime of starless cores is 0.3–1.6 Myr (3 times longer than the estimated duration of the Class 0 and Class I phases, §7.8).

The above facts do more than constrain the theory: most models developed in the past are inconsistent with the facts and need complete revision. First, however, we complete the picture by examining how they are distributed.

6.2.3   *Distributions*

Up to this point, we have treated cores as if they were isolated objects. They are, however, clustered within clouds, often lying like beads in a chain, within lower density streamers of molecular gas. Clusters and collections of globular-like cores are also found. In this sense, star formation is a regional phenomena. The core clustering in ρ Ophiuchus, displayed in Fig. 6.3, appears to be best describable as fractal.

It has been proposed that the stars we see are pre-determined by the core properties. Specifically, one can envisage that the masses of the stars are equal or proportional to the masses of the cores. We need to be more precise since stars of different mass have vastly different lifetimes. Thus, we compare the distribution in core masses to the rate at which stars of each mass are being created. The latter is quantified by a probability distribution function called the Initial Mass Function (see §12.7).

Mass determinations derived through studies of molecular line emission have led to reports that the two distributions are not the same. Many large fragments are detected, including unbound features, and so we describe the resulting number distribution as the molecular *clump* spectrum. The observations yield power law distributions mainly in the range

$$\frac{dN_{clump}}{d\log M} \propto M^{-(0.6\pm0.3)}, \tag{6.8}$$

which is essentially the same law as applies to clouds in general (Eq. 3.8): most clumps are small yet most mass is still contained in the largest entities. Note that throughout this book we try to express all distributions in terms which emphasise where most of the number, mass or energy lies. We thus separate any counts into logarithmic bins. The value where the distribution peaks is then a true characteristic measurement (later, we will process photon counts similarly, to present spectral *energy* distributions (SEDs) rather than photon *number* distributions).

Core mass distributions are estimated using submillimetre dust fluxes. Through the dust, one measures the total mass much more accurately. Compact gravitationally bound objects are selected. For ρ Ophiuchus, the distribution of number per unit mass is shown in Fig. 6.3. Perhaps three power-law fits are needed to characterise the distribution. As we shall see, this resembles the number-mass distribution for stars (§12.7). Note that the mean core mass is well below one solar mass and that there are very few high mass cores. These properties are typical of cores in other locations such as Taurus, Orion B and Serpens.

6.3 Classical Collapse Scenarios

Astronomers have been active in seeking the initial conditions for star formation. In the classical collapse picture the fundamental initial assumption is a core at rest but on the verge of collapse. The ideal situation of a uniform sphere is also often taken. The objective is to relate the static core properties to the final stellar properties by calculating the evolution.

It was soon realised that a uniform core would cool and lose thermal support. The collapse consequently becomes dynamic, possibly in gravitational free-fall. Systematic motions reach supersonic speeds. A small protostellar core then forms and residual gas accretes dynamically onto the core.

The full story, however, is not so simple. Besides the many choices of

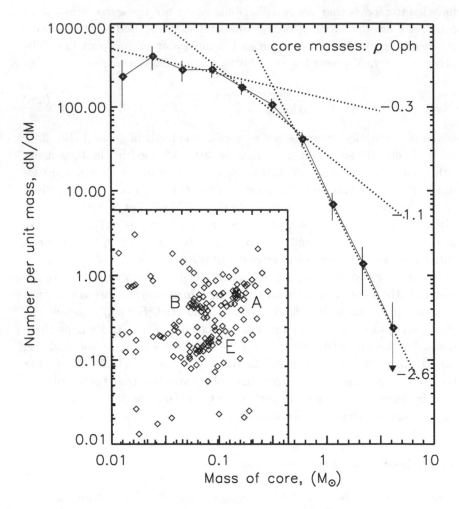

Fig. 6.3 The number distribution by mass of the cores in ρ Ophiuchus for 118 small-scale fragments. The dotted lines show power-law fits of the form $dN/dM \propto M^{\alpha}$. The error bars correspond to $\sqrt{N}$ counting statistics. The inset displays the core positions in a wide region stretching $1° \times 1.2°$, with the locations of the clumps A, B and E marked (courtesy of T. Stanke, R. Gredel, T. Khanzadyan).

initial conditions (magnetic field, rotation, shape and internal structure), there is a wide range of physics (radiation transport, self-gravity, thermodynamics and chemistry), dynamical mechanisms (magnetic field diffusion, turbulent viscosity, shock wave formation) and external influences (tidal effects, radiation field, shock waves, cloud collisions) to include.

Starless cores are cold and the long wavelength radiation emitted is free to escape. This means that, provided the temperature does remain low, we can treat the cores as isothermal. The temperature may not remain quite constant but it should be a good approximation until the collapse begins to release large amounts of gravitational potential energy. Although cold, thermal pressure remains significant: a pressure-less approximation is not really justified. The approximation, however, has provided insight, demonstrating that the free-fall collapse (in the time given by Eq. 4.1) leads to fragmentation and that the spherical symmetric solution is singular: any small deviation from symmetry will grow and the cloud will flatten.

The collapse of isothermal spheres has provided a strong focus of research. Solutions beginning from a stationary uniform cloud were first tried independently by Larson and Penston in 1969 but they remain too theoretical to exploit. Nevertheless, these solutions demonstrated the tendency towards a collapse at 3.3 times the sound speed. One goal was to find solutions in which the gas smoothly settles onto a central compact object. The initial conditions, however, first lead to a supersonic collapse which can only be abruptly braked.

In what can be considered as the standard view, Frank Shu and others have successfully argued that a singular isothermal sphere can develop with a radial distribution of the form $\rho \propto 1/R^2$. In this scenario, a core may begin in a state just beyond marginal stability, the critical Bonnor-Ebert sphere (§4.3.3). The envelopes of such states possess density profiles close to the form $\rho \propto 1/R^2$ and the central concentration becomes enhanced during an initial slow collapse, producing in theory the singular sphere. An inside-out collapse ensues in which the central region collapses and accretes well before the outer envelope. The central region then undergoes free-fall collapse and the density takes the modified form $\rho \propto 1/R^{3/2}$ and the free-fall speed is $v_{ff} \propto -1/R^{1/2}$ (since $v_{ff}^2 = -GM/R$). The size of the inner region grows as an expansion wave propagates outwards. This theory predicts that the rate at which mass falls onto a central accreting nucleus or 'protostellar core' remains constant in time since the mass inflow rate is $4\pi R^2 \rho v_{ff}$. These predictions allow the model to be tested.

6.4 Core Theory

Many origins for the cores have been proposed. Models which have found support include the following.

- Evolution of clumps through a succession of equilibrium configurations.
- Cold dense clumps in a warm diffuse medium produced through thermal instability.
- Gradual collapse through the extraction of a resisting magnetic field via ambipolar diffusion, as described in §7.5.
- Compression via long-lived non-linear Alfvén waves (§7.2).
- Compression after clump collisions.
- Chance appearance of high density peaks within supersonic turbulence.

In the complex interstellar medium, circumstances where each is responsible are likely to be met. However, we seek the dominant processes which create the majority of stars. In this sense, most models appear to be too restrictive in application.

Fragments not held together by self-gravity are the outcome of many of the above models. There are at least three processes by which a cloud fragment would then grow in mass to become a core. In addition to ambipolar diffusion of neutral material across magnetic field lines, there is Bondi-Hoyle accretion as the clump sweeps through the ambient cloud, or a cooling-driven flow onto the fragment. Other fragments may accumulate mass on a time scale which depends strongly on the mass of the fragment. One would then expect many small cores for each large core. Eventually the large cores would become gravitationally unstable and collapse to form either single or multiple star systems.

Slow infall models appear to have problems fitting the data. The classical hydrodynamic models in which a core evolves rapidly on the inside and gradually on the outside predict that a detectable protostar would be present at the core centre well before the infall regime has become extended, contrary to observations. The collapse in cores such as L1544 seems to proceed in a manner not contemplated by the standard theories of star formation. Our hard work just illustrates how little is still known about the physical conditions that precede birth.

Magnetohydrodynamic models invoke magnetically-supported cores but in which the magnetic field gradually lets the molecules slip through, in the process called ambipolar diffusion (§7.5). Lifetimes of between 3 Myr and 14 Myr are predicted although the critical parameters are not well known. Nevertheless, these lifetimes are an order of magnitude larger than our estimates based on observations.

6.5 Turbulent Evolution of Cores

The suggestion is that our natural tendency to 'begin with' something static immediately puts us on the false track. Since clouds are turbulent at suprathermal speeds, surely we should begin from a dynamic turbulent state? Yet in the observed cores, the turbulence has decayed and clearly does not dominate the pressure. Nevertheless, the 'initial conditions' in the cores are controlled by the state the gas is left in after the decay.

We begin with a turbulent cloud out of which density fluctuations generate fragments, some of which are massive enough to be self-gravitating. These fragments decouple from their turbulent environment, collapsing to protostars with little further external interaction. This mechanism may tend to form low-mass cores. High-mass stars would then require merging of several of these condensations, preferentially occurring in dense highly turbulent gas where collisions would be frequent.

The cores which form out of the turbulence are described as triaxial and distorted, often appearing extremely elongated. Computer simulations show that sheet-like structures form in a medium dominated by supersonic turbulence. The sheets collapse into filaments as self-gravity takes control. The filaments then split up into chains of elongated cores. Also, dense cores develop at locations where filaments intersect. Later in the simulations, as the turbulence dies away, the cores tend to become more regular since the physics becomes dominated by thermal pressure and gravity.

Vital themes not mentioned until now are that of binary formation, cluster formation and the differences between the formations of high and low mass stars. These are themes within which we can compare the objects that we develop against the finished products (see §11.9). Nevertheless, it should be remarked here that we will have to find the means to fragment a core at some stage, and also to form star clusters with diverse properties.

After a long delay, the work on supersonic turbulence is progressing fast and we will soon be able to predict the core structure and evolution in detail. The results will depend on the physical and chemical input into the computer simulations. Therefore, although it is clear that cores are the mediating phase between the cloud and the collapse into stars, we need to take a fresh look at the context in which the physical processes operate. This we do in the next chapter. First, however, it is worthwhile to let the core collapse run its course up to the critical moment.

6.6   The Approaching Birth

We are now preparing for the moment of birth. Final adjustments have to
be made due to the increasing density. The thermodynamics now take on
supreme importance. Much of what occurs is still theory but here is what
might well occur:

Stage 1. From the outset, the central density shields the core from
external radiation, allowing the temperature to drop slightly as the dust
grains provide efficient cooling. Therefore, the hydrogen is molecular.

Stage 2. A roughly isothermal collapse all the way from densities of
$10^4 \, \mathrm{cm}^{-3}$ to $10^{11} \, \mathrm{cm}^{-3}$ then proceeds. The gravitational energy released
goes via compression into heating the molecules. The energy is rapidly
passed on to the dust grains via collisions. The dust grains re-radiate the
energy in the millimetre range, which escapes the core. So long as the
radiation can escape, the collapse remains unhindered.

Stage 3. At densities of $\sim 10^{11} \, \mathrm{cm}^{-3}$ and within a radius of $10^{14} \, \mathrm{cm}$
the gas becomes opaque to the dust radiation even at $300 \, \mu\mathrm{m}$ (according to
Eq. 6.6). The energy released is trapped and the temperature rises. As the
temperature ascends, the opacity also ascends. The core suddenly switches
from isothermal to adiabatic.

Stage 4. The high thermal pressure is able to resist gravity and this
ends the first collapse, forming what is traditionally called the *first core* at
a density of 10^{13}–$10^{14} \, \mathrm{cm}^{-3}$ and temperature of 100–$200 \, \mathrm{K}$.

Stage 5. A shock wave forms at the outer edge of the first core. The
first core accretes from the envelope through this shock. The temperature
continues to rise until the density reaches $10^{17} \, \mathrm{cm}^{-3}$.

Stage 6. The temperature reaches $2000 \, \mathrm{K}$. Hydrogen molecules dissociate at such a high temperature if held sufficiently long. The resulting atoms
hold less energy than the molecules did (the dissociation is endothermic),
tempering the pressure rise. The consequence is the *second collapse*.

Stage 7. The molecules become exhausted and the cooling stops at the
centre of the first core. Protostellar densities of order $10^{23} \, \mathrm{cm}^{-3}$ are reached
and with temperatures of $10^4 \, \mathrm{K}$, thermal pressure brakes the collapse. This
brings a second and final protostellar core into existence. The mass of this
core may only be one per cent of the final stellar mass.

Stage 8. The first shock from Stage 5 disappears while a second inner
shock now mediates the accretion onto the protostellar core. A star is born.

Chapter 7

The Magnetic Mediation

In the rush towards the birth, we have neglected two serious complications in the pregnancy. First, cores can't just collapse because they will end up spinning too fast and flying apart. Instead, the collapse should be limited by a 'centrifugal barrier' raised by angular momentum conservation. Since the total angular momentum of an isolated cloud is conserved, the rotation speed increases as a cloud collapses. This can clearly only continue until it approaches the escape speed, so defining the barrier.

The second complication is the magnetic field. Magnetic fields are thought to play a significant role in all stages of star formation. As the collapse progresses, the trapped magnetic field is intensified and the magnetic pressure begins to resist the infall. We can envisage this as a 'magnetic barrier' which holds up star formation.

A third controversy is then raised by the turbulent motions in molecular clouds. To maintain the turbulence, we would require some means to drive the motions which avoids the strong thermal dissipation of sound waves. Magnetic waves could do the trick, absorbing kinetic energy from and later returning kinetic energy into the cloud. However, a configuration in which this situation is successful has not been found as yet. Instead, computer simulations show that energy dissipates through magnetic waves as fast as through acoustic waves.

Here, we need to uncover how nature overcomes classical problems of star formation. Magnetic field concepts are often invoked by theoreticians in a crisis, and we are required to delve into the depths of magnetohydrodynamics. To confront theory, however, we first need to know if a significant magnetic field is really present.

7.1   Magnetic Field Observations

The influence of magnetic forces is still not fully understood, in part because empirical information is scarce. The magnetic field is the most poorly measured quantity in the star formation process.

The Zeeman effect remains the only viable method for measuring magnetic field strengths. The magnetic field splits radiation into left and right circularly polarised waves with minute frequency shifts. The Zeeman splitting is proportional to the field strength. Only the component of the field strength along the line of sight, B_{los} and its sign (i.e., toward or away from us) is directly obtainable. Statistically, for a large ensemble of measurements of uniform fields oriented randomly with respect to the line of sight, we can expect an average magnetic field 2 times higher and an average magnetic energy $\sqrt{3}$ times higher.

The Zeeman effect was first detected in the 21 cm atomic hydrogen line (§2.3.1). For molecular clouds, the OH molecule is sensitive to the effect up to densities of $10^7 \, \mathrm{cm}^{-3}$, and is especially amenable through absorption lines (e.g. at 1.667 GHz). Water, which is non-paramagnetic, is difficult to employ. Radio frequency spectral lines in other molecules such as CCS and CN are now proving valuable.

Masers can enhance the detectibility of Zeeman splitting (see §2.3.1). H_2O masers are quite common, believed to arise in star formation regions as a result of shock waves driven by outflows. The magnetic field dominates the post-shock pressure. Therefore, the maser properties probe the physical nature of the shock and some faith in the interpretation is required to thus deduce the magnetic field in the pre-shock cloud.

The aim is to measure the strength of a magnetic field in terms of it influence on a molecular cloud. We try to compare magnetic pressure to thermal pressure, turbulent pressure and gravitational pressure. This may all be summarised by measuring the contribution of magnetic energy to the balance or imbalance of the virial equation or inequality (§4.4). The results are all founded on one empirical relation which expresses the magnetic field strength in terms of the cloud or clump density,

$$B_{los} \sim 0.7 \times 10^{-6} n^{0.47} \, \mathrm{G}, \tag{7.1}$$

(in units of Gauss), which may hold for densities in the range $100 < (n/\mathrm{cm}^{-3}) < 10^7$. Unfortunately, this relation is based on only a few detections of sources which may not be typical. Also, the scatter is high and it does not take into account measured upper limits.

The field strength in diffuse clouds is quite well established: line of sight fields are typically 5–10 μG. The magnetic energy density is then estimated at $B^2/(8\pi) \sim 10^{-11}$ erg cm^{-3}, which is roughly equal to the kinetic energy in the diffuse clouds with $n \sim 30$ cm^{-3} and $\Delta v \sim 6$ km s^{-1} (we take $B^2 = 3 B_{los}$). This value is consistent with Eq. 7.1 at the low end of the density range. The vital question is: what happens to the field during contraction?

7.2 Magnetohydrodynamics

We have already seen that the motions in molecular clouds are generally faster than that of sound waves. The supersonic speed is expressed as a Mach number (§5.1). The equivalent for magnetic waves is the Alfvénic Mach number or Alfvén number, which measures the speed relative to the magnetic wave speed, the Alfvén speed. This speed is determined by looking for waves which satisfy the full equation of motion, Eq. 4.3, including the Lorentz force.

The Lorentz force is derived from Maxwell's equations on taking into account that high electrical conductivity and that non-relativistic speeds are involved in star formation. This MHD term can be rewritten through a vector identity as the sum of a magnetic tension and a negative gradient of the magnetic pressure, $P_b = B^2/(8\,\pi)$, respectively:

$$\frac{1}{4\pi}(\nabla \times \mathbf{B}) \times \mathbf{B} = \frac{1}{4\pi}(\mathbf{B} \cdot \nabla)\mathbf{B} - \frac{1}{8\pi}\nabla(\mathbf{B}^2). \qquad (7.2)$$

Unlike the one type of acoustic wave, there are three different classes of MHD wave. Undertaking a linear analysis similar to how Eq. 4.15 was derived, we find three solutions representing the three classes. Firstly, non-dispersive waves move at the Alfvén speed, v_A, along the field lines or at a speed $v_A cos\phi$ when propagating at an angle ϕ to the magnetic field direction. The fluid is set in motion transverse to both the wave direction and field direction but there is no associated compression. Here,

$$v_A = \frac{B}{\sqrt{4\pi\rho}} = 1.85\left(\frac{B}{1 \times 10^{-6}\ \text{G}}\right)\left(\frac{n}{1\ \text{cm}^{-3}}\right)^{-1/2}\ \text{km s}^{-1}. \qquad (7.3)$$

Note that the tension of the magnetic field lines supports these transverse waves. Most notable, isolated Alfvén waves of any amplitude do not dissipate. The other two types of wave are magneto-acoustic. The magnetic pressure supports fast magnetosonic waves, which work in unison with the sound waves and so move at least as fast. In contrast, slow magnetosonic waves propagate slower than Alfvén waves due to counteracting thermal and magnetic pressures.

Magnetic effects dominate thermal effects in cold clouds. To show this, we evaluate a crucial parameter in any magnetohydrodynamic theory: the ratio of thermal to magnetic pressures, classically called the β parameter. Theorists generally compute simulations with a wide range of β since, with our limited knowledge of the strengths of magnetic fields, it remains a free parameter. Given the thermal pressure from Eq. 4.5, then $\beta = P_t/P_b = 2v_{th}^2/v_A^2$. If we take Eq. 7.1 at face value, then $\beta \sim 0.02$ for $10\,\mathrm{K}$ cold clouds (taking v_{th} from Eq. 3.9).

This result implies that in turbulent zones MHD waves would soak up 10 - 100 times more energy than acoustic waves. By tapping this reservoir, MHD turbulence could support a cloud for extended periods. It was always clear that supersonic turbulence would rapidly dissipate through hydrodynamic shocks. Purely hydrodynamic turbulence would be expected to shock and dissipate on approximately the free-fall timescale (§4.4). This led astrophysics to suggest that MHD waves might organise supersonic motions and significantly lengthen the dissipation timescale. The relevant speed is not the sound speed but the Alfvén speed. If the energy could be dumped into Alfvén waves, then there would be no need for dissipation since even in the non-linear regime these waves are resilient. In this case, hydromagnetic turbulence could support a cloud, providing a theoretical basis for the long apparent ages of molecular clouds and their low efficiency of star construction.

A significant recent result, however, is that the wave energy is *not* retained but dissipated through acoustic or MHD waves at the same rate at which it is pumped in. Recent supercomputer simulations demonstrate that Alfvén waves will dissipate on a timescale not much longer than the free-fall timescale. It was found that Alfvén waves couple to fast and slow magnetosonic waves. These waves steepen and form dissipative shock waves, draining away the MHD energy into thermal energy, and subsequently radiated away. Therefore, a recurring input of mechanical energy would still be necessary to maintain the turbulence. Furthermore, Alfvén waves must be outwardly propagating in order to provide cloud support against gravity; waves propagating inward would compress the cloud. There are potential sources of outwardly propagating waves such as torsional Alfvén waves, produced by transport of angular momentum outward as a core contracts and, later in the evolution, protostellar or stellar winds from star formation in the core.

The possibility that internal motions are due to MHD waves can be tested by calculating the Alfvén number of a flow. According to field measurements, the Alfvén speed is a constant, on applying Eqs. 7.1 and 7.3, with a value of $2\,\mathrm{km\,s}^{-1}$. The velocity dispersion in molecular clouds exceeds this value, but for clumps the flow could be described as trans-Alfvénic and for cores the speeds are sub-Alfvénic. Hence, the internal motions in clumps and cores could, at least in principle, be due to MHD waves. For a closer evaluation of the ability of a field to hold up a cloud we return to the pressure and virial analyses.

7.3   Magnetic Field and Flux

Under what conditions will gravitational collapse overwhelm magnetic repulsion? In 1956, Mestel and Spitzer introduced a critical mass associated with the amount of magnetic flux threaded through a self-gravitating cloud. To understand magnetic flux we need to complete the basic set of equations by adding the induction equation, which expresses the change in magnetic field at a fixed location as

$$\frac{\partial \mathbf{B}}{\partial t} = \nabla \times (\mathbf{v} \times \mathbf{B}), \qquad (7.4)$$

and the non-divergence initial condition

$$\nabla \cdot \mathbf{B} = 0. \qquad (7.5)$$

These are the means of expressing that a magnetic field can be induced and that magnetic monopoles (the equivalent of an electric charge) can't exist. The physical meaning this carries is that the magnetic field can be represented by tubes of flux frozen into the fluid. As a tube is squashed, the density increases and the magnetic field strength also, both inversely proportional to the flux tube area, A. That is, flux is conserved. In addition, material is free to move along the tubes without altering the field or flux. Therefore, during an isotropic collapse of a sphere of radius R threaded by a uniform field, we would expect the magnetic flux

$$\Phi = A \times B = \pi R^2 B \qquad (7.6)$$

to remain constant. With constant mass and uniform density, $M = (4/3\pi)R^3\rho$, flux freezing then predicts $B \propto \rho^{2/3}$. Note that this is is a somewhat steeper increase with density than indicated by the observations (Eq. 7.1).

According to magnetohydrodynamic theory, a cooling cloud will tend to contract along the flux tubes since magnetic pressure provides resistance to their squeezing. The theory therefore predicts that a magnetised cloud will take an oblate shape with the minor axis parallel to the field lines. With some contraction in each direction, the field achieves an 'hourglass' morphology on a large scale. Once rotation also supports the oblate structure, a contracting accretion disk further squeezes the 'waist' of the hourglass. Furthermore, the rotating disk twists the magnetic field lines into the azimuthal direction, in the disk plane, and so imprints a toroidal morphology.

An hourglass field shape is often observed, as deduced from the polarisation direction of dust grain emission. Fig. 7.1 displays the OMC-1 cloud measurements in the plane of the sky and Fig. 7.1 demonstrates a plausible interpretation of the structures in the line of sight.

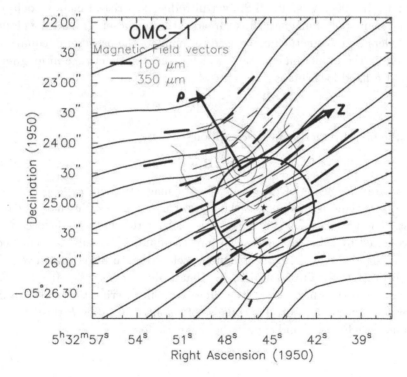

Fig. 7.1 The measured magnetic field vectors in the Orion Molecular Cloud OMC-1 region forming an hourglass shape (taken from D. Schleuning 1998, Astrophysical J., 493, 811).

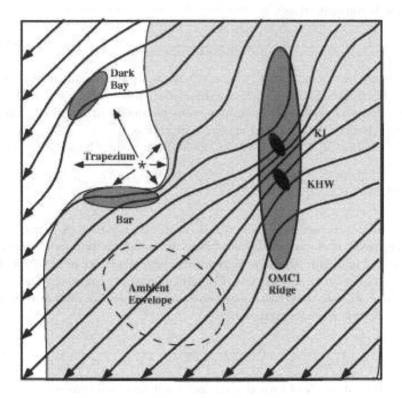

Fig. 7.2 A schematic overview of the cloud structure and magnetic field configuration in the Orion Molecular Cloud region, with the earth located 400 pc to the left of the diagram. Note the face-on appearance apart from the 'bar' which is a photo-dissociation region (taken from D. Schleuning 1998, Astrophysical J., 493, 811).

An internal magnetic field contributes to the support of a molecular cloud provided the field is tightly coupled to the gas. To examine whether the magnetic field alone can resist gravity, we apply the virial theorem (Eq. 4.4) with just the magnetic and gravitational energies and assume that these energies are in equilibrium. The magnetic energy per unit volume is taken as $B^2/(8\pi)$. This determines a critical mass corresponding to approximately equal gravitational forces and magnetic stresses:

$$M_\Phi = \frac{\sqrt{5}}{3\pi\sqrt{2}} \frac{\Phi}{G^{1/2}} \sim 0.17 \frac{\Phi}{G^{1/2}} \tag{7.7}$$

for the spherical cloud. The maximum mass which can be supported by the field assuming that the field alone opposes self-gravity, so that the cloud indeed becomes flattened, is

$$M_\Phi = \frac{\Phi}{2\pi \, G^{1/2}}. \tag{7.8}$$

These are just about equal, whereas if we took a constant mass-to-flux ratio distributed through a cloud, the critical mass can be about 0.6 of these values. It follows that an important parameter in the discussion of support is simply the mass to magnetic flux ratio.

Evaluating Eq. 7.7 (using Eq. 7.6) yields an extremely high critical mass:

$$M_\Phi = \frac{(5/2)^{3/2}}{48\pi^2} \frac{B^3}{G^{3/2}\rho^2} = 4.6 \times 10^4 \left(\frac{B}{1 \times 10^{-6} \text{ G}}\right)^3 \left(\frac{n}{1 \text{ cm}^{-3}}\right)^{-2} M_\odot. \tag{7.9}$$

It has been proposed that this critical mass, also regarded as 'a magnetic Jeans mass', is as potentially important to star formation as the Chandrasekhar limit is to stellar evolution (although the latter limit of $1.4\,M_\odot$ as the maximum for a white dwarf is rather simpler to express). The ratio of a cloud mass to the critical mass can also be written in a form convenient for evaluation from observations:

$$\frac{M}{M_\Phi} = 4.6 \times 10^4 \left(\frac{N}{10^{22} \text{ cm}^{-2}}\right)^3 \left(\frac{B}{1 \times 10^{-6} \text{ G}}\right)^{-3}, \tag{7.10}$$

where the column N of hydrogen nucleons is determined through dust continuum or molecular line emission.

If a cloud mass exceeds the magnetic critical mass, it is termed supercritical and is unstable to collapse. It may collapse in all three dimensions unimpeded by the field. The mass needed represents the size of the magnetic barrier which our classical theory faces. Notably, the critical mass is invariant under a uniform spherical collapse, raising the further problem of how the cloud could fragment into clumps and cores.

This behaviour can be perceived from a more general format by extending the virial theorem, Eq. 4.10 to include the magnetic contribution, giving the required external pressure

$$P_s = \frac{M\sigma^2}{\mathcal{V}} + \frac{b\,G}{\mathcal{V}^{4/3}}\left(M_\Phi^2 - M^2\right), \tag{7.11}$$

where $\mathcal{V}$ is the cloud volume, b is a constant of order unity and σ is the isothermal sound speed. The first two terms alone represent Boyle's law ($P\mathcal{V} = \text{constant} \times \text{temperature}$). Written like this, the magnetic forces are seen either to dilute self-gravity or to reverse it. This fact is *independent of the stage of contraction.*

In §4.3.3, we showed that by increasing the pressure on a cloud, it will finally reach a regime where gravity takes command and the collapse becomes unstoppable. The way this contraction proceeds is modified by the magnetic field. If the field is weak (super-critical cloud), a cloud which has entered the unstable regime again cannot thereafter be halted. The length scale at which this occurs is called a *modified Jeans length*. It is obviously always smaller than the Jean's length and is given by

$$\lambda_{JM} = \frac{3\,G\,M}{\pi\,\sigma^2} \left[1 - \left(\frac{M_\Phi}{M} \right)^2 \right]. \tag{7.12}$$

Therefore, if a cloud of mass M is compressed to below this size, it will collapse. If the field is strong (sub-critical cloud), however, the cloud will never enter an unstable regime. No amount of external pressure will ever induce the cloud to collapse.

In summary, clouds that are initially magnetically super-critical (unstable against gravitational collapse) will collapse on relatively short timescales, preferentially forming stars of high mass. As we have calculated, if the magnetic field is insufficient to stop the initial collapse, then its compression during collapse cannot bring the cloud into equilibrium and halt the collapse.

7.4 Super-critical or Sub-critical Collapse?

So all we need to know now is whether cores and clumps are super-critical or sub-critical. Unfortunately, the data we have positive field detections for are not decisive. Magnetic fields are certainly effective at diluting self-gravity but it remains in debate whether or not most cores lie above or below the critical mass. On the scale of clumps and cores, the kinetic and magnetic energies could be approximately the same. On the largest scales, molecular clouds and complexes are clearly highly super-critical, as has been verified by the calculation of turbulent Alfvén numbers above. The magnetic pressure in clouds is not sufficient to dominate the dynamics.

It has been claimed that cores, including those in which the field has not been strong enough to detect, are super-critical and that no case for a sub-critical core has been substantiated – there is still no hard evidence that stars form from magnetically supported cores.

One reason is that much depends on the geometry. If the clouds can be taken as initially spherical with uniform magnetic fields and densities that evolve to their final equilibrium state assuming flux freezing, then the typical cloud is magnetically super-critical. On the other hand, if the clouds are highly flattened sheets threaded by a uniform perpendicular field, then the typical cloud is approximately magnetically critical provided the field strengths for the non-detections are close to their derived upper limits. If, instead, these values are significantly lower than the upper limits, then the typical cloud is magnetically super-critical. On the other hand, if the candidate fragments in which we attempt to detect Zeeman splitting are those with well-defined high column densities, then we may be selecting against sub-critical clouds. So caution is expressed.

Theoretically, a sub-critical clump appears difficult to sustain. It would require external support to resist the high magnetic pressure from within the clump. We could consider a strong external magnetic field to be present, only a little weaker than in the clump itself. This implies that we need a high change in density across the clump boundary without a similar change in magnetic field. This implies that the Alfvén speed in the cloud far exceeds that in the clump, an effect which is not apparent in the observations.

There is a strong but speculative argument that the majority of *observed* cores should be marginally critical! All super-critical cores would have collapsed long ago and those which form now are transformed immediately into stars. On the other hand, potentially sub-critical cores would have remained as diffuse clouds, in which state they can indeed be supported by external pressure. Through this line of argument, the only objects capable of star formation at present are self-selected to take the form of marginally critical cores. This suggests that the marginally stable cores do not greatly contribute to star formation, being constantly overtaken in development by a more ambitious population of newly-forming super-critical cores. These active cores form out of the turbulent quagmire rather than from disturbances to the passive cores.

7.5 Sub-critical Contraction: Ambipolar Diffusion

The barrier to star formation, raised by the high critical mass, is surmounted if the magnetic field can be separated from the gas. This is achieved in a scenario which has subsequently been termed the 'standard picture'. The process is called *ambipolar diffusion*. In astrophysics, this is the name given to the drift of neutrals with respect to the charged particles. The magnetic field interacts directly only with the ions, electrons and charged grains and, therefore, may also drift through the neutrals. In the interstellar medium, the number of charged particles suffices to maintain strong magnetic-neutral ties and ambipolar diffusion is negligible.

However, in dense molecular clouds, ambipolar diffusion is important because there are only a few ions to provide the coupling, as already determined (Eq. 2.2). Therefore, the ionisation degree plays a key role in determining the initial conditions that precede the collapse to form a star.

In a sub-critical cloud, in which the magnetic field threading the dense gas is sufficiently large to prevent immediate collapse, the neutral molecules drift into the cores without a significant increase in the magnetic flux. The increase in core density implies that the critical mass is falling. Eventually the mass to magnetic flux ratio in the core rises to the super-critical level: dynamical collapse and star formation takes over. The envelope remains essentially in place while the super-critical core collapses.

The timescale for ambipolar diffusion is proportional to the ionisation, and it is therefore critical to develop methods of estimating the ionisation level. The abundances of various species can place tight limits in the dense gas. The question posed is: will the diffusion occur fast enough to accelerate or even permit star formation? To answer this, we must treat two distinct but interacting fluids. First, we apply the equation of motion, Eq. 4.3, to just the ion-magnetic fluid but with an extra term to account for the frictional or drag acceleration of the neutrals on the ions. We supplement this with a separate equation of motion for the neutrals which includes the opposite drag force of the ions on the neutrals.

The neutrals stream through the ions, accelerated by gravity. A drift speed $v_i - v_n$ implies an average drag force on a neutral of

$$F_{in} \propto c_a n_i(v_i - v_n), \tag{7.13}$$

where n_i is the number of ions and the detailed collision physics are included in $c_a = m_r < \sigma_{in} v >_{in}$. Here, m_r is the reduced mass (which enters from applying momentum conservation) and is given by $m_r = m_n m_i / (m_n + m_i)$ and m_i and m_n are the average ion and neutral mass. We take here $m_i = 10\ m_n$. The collisional cross-section in low-speed ion-neutral collisions will be inversely proportional to the relative speed. We thus take a constant value for $< \sigma_{in} v >_{in} = 1.9 \times 10^{-9}$.

In contrast, the few ions only feel a very weak gravitational pull but experience the full effect of the Lorentz force. Their inertia is low, which leaves a balance between the Lorentz force and the drag in their equation of motion:

$$\frac{1}{4\pi}(\nabla \times \mathbf{B}) \times \mathbf{B} = n_n F_{in}, \qquad (7.14)$$

where n_n is the number of neutrals.

The timescale for ambipolar diffusion to separate ions and neutrals, originally occupying the same space, by a distance D is written as $\tau_{AD} = D/(v_i - v_n)$ where $v_i - v_n$ is given from Eq. 7.14. For simplicity we take $(\nabla \times \mathbf{B}) \times \mathbf{B} = B^2/D$ to yield

$$\tau_{AD} \sim \frac{4\pi \, c_a \, n_i n_n \, D^2}{B^2}. \qquad (7.15)$$

In terms of what may be typical core parameters with a strong magnetic field this is

$$\tau_{AD} \sim 34.2\left(\frac{n_i}{10^{-7} \, n}\right)\left(\frac{n}{10^5 \text{ cm}^{-3}}\right)^2\left(\frac{D}{1 \text{ pc}}\right)^2\left(\frac{B}{4 \times 10^{-4} \text{ G}}\right)^{-2} \text{ Myr.} \qquad (7.16)$$

This timescale is 10–20 times longer than the star formation times already discussed in §6.2.2. There is, however, considerable uncertainty in the parameters and there have been claims that ambipolar diffusion is still a viable mechanism.

We test this by substituting for the field from Eq. 7.9 to obtain a very neat and decisive formula:

$$\tau_{AD} \sim 14.0\left(\frac{n_i}{10^{-7} \, n}\right)\left(\frac{M}{M_\Phi}\right)^{2/3} \text{ Myr.} \qquad (7.17)$$

This implies that the only way a cloud can collapse fast when critical is if the ion fraction is extraordinarily low. Clouds sufficiently subcritical to collapse would, instead, expand (which explains their absence observationally).

In term of the free-fall time given by Eq. 4.1, this becomes

$$\frac{\tau_{AD}}{\tau_{ff}} \sim 101\left(\frac{n_i}{10^{-7} \, n}\right)\left(\frac{n}{10^5 \text{ cm}^{-3}}\right)^{1/2}\left(\frac{M}{M_\Phi}\right)^{2/3}, \qquad (7.18)$$

and on substituting for the ion fraction from Eq. 2.2, we finally obtain

$$\frac{\tau_{AD}}{\tau_{ff}} \sim 40\left(\frac{M}{M_\Phi}\right)^{2/3}, \qquad (7.19)$$

independent of all other parameters.

It is clear that ambipolar diffusion is not viable as a general star formation mechanism *under these standard conditions*. Where observed, large inward speeds (up to $0.1 \, \text{km} \, \text{s}^{-1}$) are too fast to result from ambipolar diffusion. The process may be critical to the observed starless cores, those self-selecting long-lived cores which do not exhibit fast infall.

The argument is, however, far from over. If the fractional ionisation is below the cosmic-ray induced level then we can revive the ambipolar diffusion scenario. This may occur deep within a core. For example, in the high-density depleted nucleus of L 1544, models do not uniquely define the chemistry but they do suggest that the ionisation level is down to a few times 10^{-9} (within 1000 AU). The ambipolar diffusion timescale is then comparable to the free-fall time for this very low level of ionisation and the nucleus is rapidly developing toward a situation where it will collapse with ionised species moving with speeds similar to those of the neutrals. A result of this argument, however, is that the cores are simply dynamical entities and a dynamical model is relevant.

Nevertheless, the timescale for ambipolar diffusion is proportional to the degree of ionisation in the core as a whole. To evaluate this mechanism, it is therefore crucial to estimate the ionisation degree based on the abundances of various species that trace the ionisation throughout the gas. Calculated values within the extended core even of L1544 exceed 10^{-8}, implying that ambipolar diffusion is inefficient.

The flat inner radial density profiles of starless cores are consistent with ambipolar diffusion contraction. The profiles are reminiscent of the quasi-static structures expected in hydrostatic equilibrium such as Bonnor-Ebert spheres. Observations of extinction through one dark isolated core, B68, show impressive agreement with the distribution expected for such a sphere. However, it is clear that reality is more complicated since the contours of dust emission/absorption as a rule are far from being spherically symmetric.

Ambipolar diffusion is a significant process in star formation but exactly what it is responsible for remains uncertain. Ion-neutral drift could be significantly faster in a turbulent medium than in a quiescent one since the turbulent motions could amplify magnetic pressure gradients. Moreover, ions and neutrals may drift in an instable manner leading to filamentary or sheet-like regions which are essentially evacuated by the neutrals. The magnetic field lines within a collapsing sheet may then reconnect generating a concentrated release of energy.

There is another problem which ambipolar diffusion may go a long way toward resolving. This is the magnetic flux problem: there are several orders of magnitude discrepancy between the empirical upper limit on the magnetic flux of a protostar and the flux associated with the corresponding mass in the pre-collapse core. Therefore, it is believed that ambipolar diffusion is eventually 'revitalised' deep in the nucleus within a core. After a protostar starts to grow in the centre, a decoupling of the magnetic flux from the inflowing gas occurs. A 'decoupling front' propagates outward within which decoupling has occurred. The front takes the form of a hydromagnetic shock.

Other models for star formation have been suggested in which the ion level plays a pivotal role. Photo-ionisation could regulate star formation since the ambipolar diffusion rate depends so strongly on the ionisation rate. Such conditions might arise in the outer regions of cores ionised primarily by UV photons whereas the inner regions are ionised primarily by cosmic rays. Confrontation of theoretical results with observations is essential for an evaluation of the many possibilities sketched above and their conflicting conclusions.

7.6 Spin

We first test the hypothesis that the angular momenta of star-forming clouds originate from the rotation of the Galaxy. In favour, the spin axes of Giant Molecular Clouds tend to be aligned with that of the Galactic disk. Their spin may thus be derived from the orbital rotation of the gas in the part of the disk where the clouds form. However, the observed direction of the angular momentum vectors can be either prograde or retrograde (parallel or antiparallel) to the spin of the Galaxy. Prograde rotation would be expected if the rotation curve of the Galaxy (rotation speed as a function of Galactic radius) is rising. The effect of vorticity generated from the interaction of the interstellar gas with large-scale spiral shock fronts can contribute significantly to the spin of giant molecular clouds. In addition, large-scale turbulence caused by supernovae within the disk will contribute. These effects, along with the details of the geometric configuration, are probably responsible for yielding retrograde rotation of giant molecular clouds in areas where the rotation curve of the Galaxy is rising or flat.

Continuing in the above fashion, sub-regions would derive their angular momentum from their parent clouds. The first angular momentum problem is that sub-regions, as well as resulting binary orbits and stellar rotation axes, show no preferred direction. Their spin axes are randomly orientated.

Before we understand how a star can form from a cloud, we must resolve a second angular momentum problem. The angular momentum of observed GMCs is indeed high, as shown in Table 3.5 with rotation periods $2\pi/\Omega$, of order of 100 Myr consistent with Galactic rotation. The problem arises in the subsequent evolution because the angular momentum per unit mass which remains within a dense core is orders of magnitude smaller. Where has the angular momentum gone? If a fragment of a cloud were to contract in isolation then the average specific angular momentum should have remained unchanged. Furthermore, the specific angular momentum of a dense core can be orders of magnitude greater than that of the single or binary stars which ultimately form (but not always, see Table 3.5). Consequently, there must exist at least one mechanism which transfers the angular momentum from a contracting fragment into the surroundings over a wide range of scales.

In standard theory, there is a quasi-static contraction phase during which cores evidently lose angular momentum by means of magnetic braking. Braking occurs via the transfer of angular momentum to the ambient gas through magnetic torques applied by torsional Alfvén waves. The process generally tends to align the angular velocity and large-scale magnetic field vectors of nearby cores. In support, well-ordered magnetic fields are observed with significant deformation from a straight parallel configuration only in the innermost regions.

To see how magnetic braking functions, we consider a rotating cloud within a static medium. The magnetic field is represented by field lines to which the gas is frozen. The frozen-in assumption holds, provided the ambipolar diffusion timescale far exceeds the rotation timescale. The field lines connect the cloud to the ambient medium. The differential motion at the interface will perturb the magnetic field, generating Alfvén waves which are likely to propagate efficiently into both media since they are (at least under ideal conditions) non-dispersive. These waves are helical in geometry on taking an axisymmetric configuration. For this reason, they are described as torsional Alfvén waves.

Firstly, consider the case where the field lines lie in transverse directions to the cloud spin axis. This is called the *perpendicular rotator*. The field lines which extend through the cloud and into the surroundings are wound up as the cloud spins. Being frozen together, the surrounding gas must be set into motion. The result is that the surroundings begin to spin as the field tries to twist the gas into solid body rotation. The moment of inertia (see §4.3.1) of the ambient gas, however, strongly opposes this action, transmitting Alfvén waves back into the cloud, braking the cloud. The braking is highly effective since the ambient material which absorbs the spin is distant from the cloud spin axis. That is, the ambient gas possesses a high moment of inertia. The timescale for the transfer of angular momentum is essentially equal to the time it takes for torsional Alfvén waves to propagate away from a core and set into motion an amount of external matter with moment of inertia equal to that of the core.

A parallel rotator is not so easily braked. The field lines extend along the spin axis through the cloud and into the ambient medium. The field lines take on a helical configuration through the spin, effected by the torsional Alfvén waves. Therefore, the surrounding material, set in motion by the spin, has a relatively low moment of inertia. To brake the cloud, much more external material must be disturbed, requiring that the Alfvén waves travel farther. As a result, we may expect magnetic fields to evolve toward being aligned with the spin axes of the local clouds. Oblique rotators tend to become aligned rotators. Furthermore, neighbouring clouds may be magnetically linked and so also evolve toward being aligned rotators.

We can estimate the timescale, t_b, for a cloud permeated by a uniform parallel magnetic field to lose its angular momentum. It is evidently proportional to the column density of the cloud along the field lines, N_p, and inversely proportional to the external particle density and Alfvén speed, n_e and v_{ae}, in the surrounding medium. We find

$$t_b = \frac{N_p}{2n_e v_{ae}}. \qquad (7.20)$$

Choosing typical parameters:

$$t_b = 7 \left(\frac{N_p}{2 \ 10^{22} \ \text{cm}^{-2}} \right) \left(\frac{n_e}{10^2 \ \text{cm}^{-3}} \right)^{-1} \left(\frac{v_{ae}}{5 \ \text{km s}^{-1}} \right)^{-1} \text{Myr}. \qquad (7.21)$$

The estimated loss rate is thus sensitive to the ambient density and slow only in low mass clouds. Higher loss rates are probable as the magnetic field fans out from the cloud into the ambient gas. In any case, it is anticipated that the sub-critical contraction of a large cloud is delayed until co-rotation of the cloud and ambient gas is attained. The centrifugal force then has little subsequent effect on the collapse.

Also in standard theory, once dynamical collapse is initiated and a core goes into a near free-fall state, the Alfvén waves are trapped. After the production of a super-critical core, the specific angular momentum is expected to be approximately conserved. This results in a progressive increase in the centrifugal force that eventually halts the collapse and gives rise to a rotationally supported disk. The subsequent evolution is described in §9.4.2. Subsequently, ambipolar diffusion may allow the material to slip across the field. One idea is that, before this stage is reached, there is a critical density where magnetic braking becomes ineffective and this could determine the binary periods of stellar systems.

We expect magnetic braking to be highly effective. Observationally, we have established that the majority of dense cloud cores show evidence of rotation, with angular velocities 10^{-13} s^{-1} that tend to be uniform on scales of 0.1 pc, and with specific angular momenta in the range 4×10^{20}– 3×10^{22} cm^2 s^{-1}, orders of magnitude lower than contained in lower density structures (Table 3.5). Although present, the rotation contributes no more than a few percent to the dynamical support of these cores. Self-gravity is mostly balanced by magnetic and thermal stresses. The ratio of rotational kinetic energy to gravitational energy lies between 4×10^{-4} and 0.07, indicating that rotation is not energetically dominant in the support of cores.

7.7 MHD Turbulence

The widths of emission lines in molecular clouds imply supersonic motions. The line shapes and their variation in space indicate an irregular velocity field rather than uniform rotation. The small rotational component in clouds may derive from the vorticity associated with the turbulence. In the same manner that momentum can be transferred at speeds exceeding the Alfvén speed, torsional MHD waves can transfer angular momentum at high speed within a turbulent medium. Furthermore, in simulations of turbulent cores, systematic velocity gradients in the line-of-sight components of the velocity are present in many cases. This is found even though the motions are actually completely random. The values derived are in good agreement with the observations.

The role of the magnetic field in cloud support has been questioned since MHD simulations indicate that unforced MHD turbulence decays in about a free-fall time, and since examination of stellar ages indicates that star-forming molecular clouds may not require long-lived support from either turbulence or magnetic fields.

However, what happens when the turbulence decays will depend on whether the quiescent remnant is magnetically super-critical or sub-critical.

In the first case, the core then collapses on the gravitational collapse time scale. Otherwise, we would be left with a magnetically supported core. Without pressure confinement, however, this core would then disperse, driven out by its own high magnetic pressure.

7.8 Summary

It is extremely important to determine whether clouds are generally magnetically sub-critical (slow evolution, dominated by ambipolar diffusion) or super-critical (fast evolution, dominated by collapse and fragmentation).

It has long been thought that magnetic braking and ambipolar diffusion are the two dominant processes in the slow formation and contraction of cores.

At present, the evidence points towards clouds being super-critical, with most clouds that are observed being those self-selected as close to marginally critical. The magnetic barrier is then overcome since magnetic pressure is unable to suppress a collapse at any stage even when frozen into the gas.

Ambipolar diffusion operates within cores to separate contracting gas from magnetic field but it is probably too slow to generate core collapse on its own. In combination with turbulence, the diffusion may be amplified. Ambipolar diffusion will also be invoked to interpret shock waves and may also be critical within accretion disks.

Magnetohydrodynamic turbulence would appear to control star formation much as described for hydrodynamic turbulence in molecular clouds (§4.4) and in starless cores (§6.5). Magnetic waves provide some extra support for the turbulence but do not alter the conclusions.

Chapter 8

The Birth

The moment of star birth is here *defined* as the instant at which a final static core has come into existence. Its mass is tiny in comparison to the final mass it will achieve. It is merely the central object, the baby, which will accrue material from the nurturing core. The moment probably corresponds to the instant at which there is no going back; the formation of a stellar system or star is inevitable. If the baby is not adequately fed, however, it may have to settle to be a brown dwarf. Or, if essentially starved, a lonely planet.

In this chapter, we describe the youngest protostellar stages and discuss the processes and problems which nature presents. First, however, we must determine if we really have observed a birth. As always, we are led by the observations and need to sort out the observational signatures which reveal how long a protostar has been alive. The interpretation remains challenging and problematic since we cannot directly observe the processes. In addition, evolutionary models don't quite succeed.

We continue our stellar biography from the proposed moment of birth. In so doing, we reverse the historical path of discovery which led from the adolescent young stars which were optically visible (Class III) back in evolution to the protostars (Class I). The holy grail was later uncovered in what was then labelled as the Class 0 protostar: a core which does not contain a Class I protostar, yet in which a collapse has brought into being a central object. To present our epic journey from cloud to star, we introduce the objects in the reverse order to their discovery.

The signatures and evolution are significantly different for stars which end up like our Sun (low-mass stars) and hot OB stars (high-mass stars). We concentrate on the birth of low mass stars here and the formation of high mass stars in §10.10.

If in the Class 0 protostar we have truly uncovered the defining moment, then we have unified our evolutionary studies. One set of astronomers were attempting to follow the evolution of molecular clouds, discovering regions of increasing compactness. Other groups were tracing back the evolutionary tracks of young stars. Now, we could link the two together and present a continuous evolution all the way through. Things are never quite that simple of course. The new complexity is that the evolution is rapid: clouds and young stars evolve simultaneously in the same region. They also interact, influencing or even regulating each other's evolution. Finally, it must be questioned whether the search is really over or whether there are objects and stages still to be discovered.

8.1   Commencement of Life

The critical moment is called *Age Zero* by some. It is actually the point at which a protostar rather than a star is brought to life. The protostar is contained within a protostellar core, distinguished from the 'starless cores' (actually 'protostar-less-cores') described in the previous chapter. One could define a star's life to begin at other critical moments, raising similar controversies to that surrounding the beginning of human life. We could instead choose the core itself as the baby, or wait until hydrogen nuclear burning has begun. By analogy, however, cores are commonly treated like eggs while nuclear burning represents the approach towards adulthood.

Age Zero should help define a unique clock for each star. We take it to correspond to the moment when a gaseous sphere is thermally enclosed for the first time. Enclosure is reached in the state where photons cannot directly escape from a sphere but are radiated from a photosphere. There is a bottleneck of photons within the photosphere: the energy reservoir inside cannot be efficiently emptied. This concept of birth corresponds to Stage 3 of the approach to birth discussed in §6.6, at the end of the first collapse. Nevertheless, this event is only about 2,000 years before the final collapse of Stage 7.

A rather subjective date of birth is also associated with the instant at which a young star is exposed to us in the optical or visible part of the spectrum. After this, the object can be treated with the traditional tools developed for stellar evolution. That means we can find its temperature and luminosity and so, from then on until its death, predict its evolution as a track on the standard Luminosity-Temperature diagram (Fig. 9.2) or 'Hertzsprung-Russell' diagram (Fig. 9.2), as described below. A 'birthline' is then expected. This line is the predicted locus of points on the diagram which divides an empty region (obscured protostars) from optically-detectable young stars of all possible masses.

8.2 Identifying and Classifying Protostars

How can we determine the evolutionary status of a star that we can't even see? The status of a star in classical astronomy is determined by locating it on a Hertzsprung-Russell diagram (Fig. 9.2). This requires measurement of a star's luminosity and colour. The luminosity is determined from its observed flux and distance. The colour is closely related to the wavelength at which the emission peaks which is determined by the effective surface temperature if it is a blackbody. The surface temperature can also be deduced by measuring atomic spectral lines and applying some atomic physics since many lines are sensitive indicators of the temperature. Visible stars are therefore also classified in terms of luminosity class and spectral type.

The trouble is that young stars and protostars are not classical stars, as is clear from Fig. 8.1. At optical wavelengths, the youngest stars are rendered invisible. They are often not blackbodies and the spectral type is often absent or ambiguous. For the deeply embedded stars discussed in this chapter we only see light which has been processed by the enveloping core out of which they grow. Gas and dust absorb the protostar's radiation and re-radiate it at much longer wavelengths. Even at infrared wavelengths, such protostars can be hard to detect. As a further complication, the escaping radiation from the circumstellar material usually arises from a wide range of distances from the central source, generating an extended continuous spectrum which does not resemble a blackbody.

Nevertheless, we can still determine the total luminosity escaping from the envelope given enough observations spread over the wide continuum. Such broad-band photometric observations are crucial for the understanding of protostars. The resultant radiated power is called the bolometric luminosity.

We also need an appropriate substitute for the optical colour. Two possible substitutes have been seriously considered. The established method was originally proposed by Charles Lada and Bruce Wilking in 1984. It involves the properties of the *spectral energy distribution*, or SED, the distribution of radiated power with wavelength. Rather than following the radiated power per unit wavelength, S_λ, we study the distribution of $\lambda \times S_\lambda$.

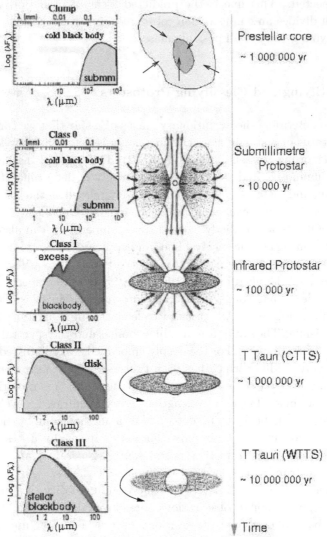

Fig. 8.1 A schematic evolutionary sequence for the SED and components from embedded clumps to naked young stars. (Credit: adapted from C. Lada, P. André, M. Barsony, D. Ward-Thompson.)

More precisely, we calculate a property of the SED measured over a narrow range in wavelengths: the infrared SEDs. The earliest protostars should be still surrounded by massive infalling envelopes which radiate copiously in the far-infrared. In contrast, more advanced stages, where the envelope has been incorporated into the star or dispersed, should radiate strongly in the near-infrared and optical. Therefore, the slope of the SED in the infrared ought to provide an empirical evolutionary sequence.

Four broad classes of young stellar object are usefully defined. These are designated as Class 0, I, II and III and illustrated in Fig. 8.1. In this chapter, we restrict our attention to the Class 0 and Class I sources which peak in the submillimetre and far-infrared portions of the spectrum, respectively. Whereas S_λ gives the luminosity within each unit wavelength interval, $\lambda \times S_\lambda$ is a quantity which emphasizes the dominating spectral region. To obtain a spectral index which describes the form of the power-law function, we take the logarithm and fit a straight line. The slope of this line α_λ across the wavelength range 2–20 μm defines an infrared spectral index:

$$\alpha_\lambda = \frac{d(\log \lambda S_\lambda)}{d(\log \lambda)}. \tag{8.1}$$

Often, it is more useful to employ the frequency as the variable:

$$\alpha_\nu = \frac{d(\log \nu S_\nu)}{d(\log \nu)}, \tag{8.2}$$

in which case $\alpha_\nu = -\alpha_\lambda$.

Class 0 and I are then defined as possessing infrared SEDs that rise with increasing wavelength. That is, the spectral index is positive ($\alpha_\lambda > 0$). They are sources thus identified as young protostars, deriving most of their luminosity from accretion, embedded in a massive envelope. More precisely, we will use the terms Class 0 and Class I *sources* to describe the types of cores which contains the Class 0 and Class I *protostar* in their interior. respectively.

To divide Class 0 from Class I protostars, we require a further empirical border based on how we expect a protostar to evolve. We have seen that most of the luminosity of a low-mass core is emitted at infrared wavelengths because higher energy radiation from the central protostar is reprocessed by dust grains in the envelope. These heated grains also emit detectable radiation at submillimetre wavelengths. This cold dust emission samples the entire mass contained within the extended envelope because the envelope is optically thin at such long wavelengths. Class 0 sources are defined to be cores which possess a central protostar, but which have $L_{smm}/L_{bol} > 0.005$, where L_{smm} is the luminosity at wavelengths exceeding 350μm. Clearly, if we presume that the quantity L_{smm}/L_{bol} will only decrease with time, as the cool envelope fades, then this definition should pick out the youngest protostars.

The second colour substitute is the bolometric temperature. This entails numerous measurements over a wide range of wavelengths with a range of telescopes, measurements which are now becoming increasingly achievable. When accomplished, the bolometric temperature may be a guide to the phase in the growth of the protostar, increasing with age. For a blackbody, the wavelength range within which the quantity $\lambda \times S_\lambda$ peaks automatically yields the bolometric temperature, as defined by Eq. 6.7.

Fortunately, so far there has been no serious contradiction between the two systems. Quite sharp divisions in the bolometric temperature correspond to the class divisions. Class 0 corresponds to protostars with $T_{bol} < 70 \, \mathrm{K}$ and Class 1 to $70 \, \mathrm{K} < T_{bol} < 650 \, \mathrm{K}$. To see how closely these classes correspond to a physical classification, we require an interpretation. This, of course, requires a model. However, even if we assume a systematic development, the bolometric temperature is more likely to be controlled by the luminosity rather than the age. Therefore, until we have more data, interpretations remain open.

8.3   Observations of Protostellar Cores

In the past, only single bolometers were available to detect the emission from the cores. Now, high sensitivity submillimetre bolometric arrays, such as SCUBA, allow us to map the dust emission from the envelopes. The resulting picture is still being pieced together.

We are finding that cores which have successfully given birth appear generally very different from starless cores. They reveal their new status by taking a more compact shape. Starless cores tend to have a flat density profile at small radii and are bounded or sharp-edged at some outer radius, thus resembling finite-sized Bonnor-Ebert spheres. In contrast, the density in protostellar cores is more centrally peaked with a density profile corresponding closer to the form $\rho \propto R^{-2}$.

Furthermore, the velocity structure is distinctive. Infall signatures are much more obvious and they exhibit broader lines attributed to substantial turbulence. The origin of this turbulence is thought to be the protostar itself which feeds kinetic energy back into the core not only through the release of gravitational energy in the form of radiation, but also through jets in the form of kinetic energy (see §9.8). Typical values for the width of NH_3 lines are 0.4 km s^{-1} in protostellar cores as opposed to 0.3 km s^{-1} in starless cores. Similarly, the average width of $C^{18}O$ lines are 0.6 km s^{-1} in protostellar cores as opposed to 0.5 km s^{-1} in starless cores.

Although the cores are gravitationally bound, they are not usually observed in isolation. What spatial scales of core clustering are set during star formation? How does a clump fragment into star-forming cores as it contracts? To answer these questions we have begun to trace the number, relative position and extent of cores. Molecules with large dipole moments such as CS, H_2CO (formaldehyde) and HCN (hydrogen cyanide) are good tracers of the density because of the large volume density required for thermal collisional excitation. Symmetric top molecules such as methyl cyanide (CH_3CN) are good tracers of temperature.

Three morphological types of cores are distinguished observationally: independent envelope, common envelope and common disk systems. The independent envelopes are separated by over 6,000 AU. Common envelope systems consist of one large-scale core which splits into multiple components on the scale 150–3,000 AU. Common disk systems lie within the same disk-like structure on scales of $\sim$ 100 AU. These structures must be related to the nature of the resulting stellar system. This requires an understanding of the fragmentation processes which we examine in §12.5. Here, we focus on the individual cores and the primary protostar growing within.

8.3.1 Class 0 protostars: the observations

Class 0 sources are surrounded by large amounts of circumstellar material. They are so highly enshrouded that their SEDs peak longward of 100 μm and their near-infrared emission is very faint. Most are not even detected shortward of 20 μm. One example of a Class 0 SED is displayed in Fig. 6.1. The spectral widths are quite similar to those of single temperature blackbody functions with temperatures in the range 20–70 K, warmer than starless cores.

All Class 0 sources are associated with molecular bipolar outflows. The outflows are high-powered and well-collimated. These will be described in §9.8. Observationally, they are signposts for otherwise hidden protostars. Physically, they may prove to be a necessary consequence of a process essential for the final collapse: channels for removing excess angular momentum.

8.3.2 *Class I protostars: the observations*

Class I SEDs are broader than predicted by single blackbodies and peak in the far-infrared. Nevertheless, we can associate bolometric temperatures through Eq. 6.7. We find that the range $70\,\mathrm{K} < T_{bol} < 650\,\mathrm{K}$ covers Class I objects. There is a huge 'infrared excess' and they often exhibit a sharp absorption feature at 10μm due to silicate dust. Hence, a large amount of warm circumstellar dust generates the infrared properties.

Although obscured from view in the optical, atomic emission lines are observed in the infrared. Apart from this, the infrared is virtually feature-less and heavily veiled (i.e. the continuum dominates at all wavelengths). A significant fraction of the near-infrared emission is scattered light, detected as a small near-infrared reflection nebula.

Class I sources are also often associated with molecular bipolar outflows. They are typically less energetic than the Class 0 outflows but the vast majority of detected molecular outflows are driven from Class 1 sources.

It is not clear how the luminosity evolves. Class I protostars are not significantly less luminous than Class 0 protostars. The luminosity range is ~ 0.5–50 $\mathrm{L}_\odot$. They are however, more luminous than the older Class II sources in the ρ Ophiuchus region but not in the Taurus region.

How can we be sure that these sources are protostars and not some other type of star which happen to be embedded? There are several pieces of evidence:

- They are located within star-forming clouds.
- Modelling in terms of a rotating and infalling cloud core is able to reproduce the features of the SEDs.
- Inferred mass infall rates are of order 10^{-5} $\mathrm{M}_\odot$ yr^{-1} which implies that a solar mass star could form in 100,000 years. This is consistent with their statistically determined lifetime.
- The infrared excess indicates that there is plenty of heated dust close to the source. This dust would be blown away by radiation pressure and would not survive in a steady state in this region. Therefore, a dynamically infalling envelope is suggested.
- Direct kinematic evidence for infall has now been observed in a number of Class 0 protostars. The signature is an asymmetry in the line profiles in which the red-shifted portion of an optically thick emission line is depressed relative to the corresponding blue-shifted portion. In fact, the first example of this signature was that of IRAS 16293-2422, observed in transitions of CS.

8.4 Theory of Accretion onto Core

Figure 8.1 shows schematic pictures of the main components of protostellar systems: the central protostar, the circumstellar disk, and the surrounding envelope. The observed Classes are interpreted in terms of these three physical components as follows.

The central protostar is classified as a Class 0 protostar as long as $L_{smm}/L_{bol} > 0.005$. This was originally meant to correspond to the case where more than half of the total mass of the system is still in the infalling envelope. After half of the mass of the envelope has fallen in, the source is referred to as a Class I protostar. This, however, depends on the chosen model for the mass–luminosity relationship.

Once almost all of the envelope has been accreted (or otherwise dissipated), the source is referred to as a Class II source or Classical T Tauri star (CTTS). At this stage, there is still a substantial circumstellar disk, which may go on to form planets.

Finally, when the inner part of the disk has dispersed, it is known as a Class III source or a weak-line T Tauri star (WTTS).

The processes by which material moves between the components, from the outer envelope to the central protostar, are still under debate. Theories differ primarily in their predictions of the density, $\rho(r, t)$, and velocity, $v(r, t)$, as a function of radius, r, and time, t. Cores are, however, as a rule not uniform or spherical and the observed quantities are angle-averaged. The three broad categories of theory are founded on:

(1) the loss of gravitational stability and collapse of a static core,

(2) the triggered compression and collapse of a static core and

(3) the gravitational collapse of a fragment produced dynamically through turbulence.

In other words, the interpretation of protostellar cores depends on the chosen *initial conditions*. Once we know $\rho(r, t)$ and $v(r, t)$, the predicted evolutions can be compared through the mass accretion rates given by

$$\dot{M}(r, t) = -4\pi r^2 \rho(r, t) v(r, t). \tag{8.3}$$

We can calculate how the bolometric luminosity evolves from the energy released during the accretion. Protostellar sources begin by being gravitationally powered; the nuclear power from the protostar is low. We assume that the radiation escapes on a timescale that is short in comparison to the dynamical timescale and that the energy leaks out (or is stored) in no other form. Then, the gravitational energy released per unit accreted mass is approximately GM_*/R_* for a protostar which has reached a mass M_* and radial size R_*. This is because most of the energy is liberated abruptly at an accretion shock near the protostellar surface. The accretion shock may take the forms of a spherical shock and an equatorial shock, according to how the material approaches. That is, we simply take the potential energy from infinity to the protostar's surface. This energy multiplied by the rate at which mass accretes $\dot{M}(R_*, t)$ onto the surface yields the total energy released, called the accretion luminosity:

$$L_{bol}(t) = \frac{G\dot{M}M_*}{R_*}. \tag{8.4}$$

This formula will be more accurate, the more the gravitational energy is released from close to the protostellar radius and the closer the accretion is to being uniform. However, some gravitational energy is obviously released during the initial collapse phase before protostar formation. This energy is transferred into the gas via compressional heating and could be detected in the future.

To determine L_{bol} we also need to know how the protostar is growing. While $M_*(t)$ corresponds to $\dot{M}$ integrated from Age Zero up to time t, the radius $R_*(t)$ requires some knowledge of the density in the protostellar interior. Moreover, there are more assumptions needed. For example, some of the mass may be diverted into a secondary protostar or smaller objects, and some may be diverted into an outflowing wind or jets. Despite these unknowns, astronomers are irrepressible and we expect to eventually take these factors into account.

8.5   Accretion Rates from Static Initial States

Initial conditions of uniform density and zero velocity were examined by Larson and Penston in 1969. This classical model encounters problems due to its prediction of supersonic infall velocities that are not observed in regions of low-mass star formation. Perhaps a more consistent static initial condition is a Bonnor-Ebert sphere in which a uniform density is approached only in a central region. Computationally, an equilibrium sphere is set up and then perturbed. Hydrodynamic calculations of the collapse subsequent to protostar formation show that the density approaches a R^{-2} radial distribution at small radii. The accretion rate in this case is not constant in time but rises sharply and then falls gradually.

There are just two fundamental parameters determining accretion rates: the speed of sound, c_s, and the gravitational constant, G. This is, in essence, because thermal pressure gradients balance gravity ($M \sim \beta c_s^2 R/G$) and the timescale is $t \sim R/c_s$ Together, they yield a mass rate of

$$\dot{M} \sim \beta \frac{c_s^3}{G},\qquad(8.5)$$

where β is a constant of order unity.

In 1977, Shu proposed a scenario where infall begins from the inside, i.e., an inside-out collapse. The initial density configuration is a singular isothermal sphere with a $\rho(R) \propto R^{-2}$ power law. The collapse region begins at the centre and is bounded by the front of an expansion wave which propagates outward at the sound speed, c_s. Outside this wave, the cloud core is at rest and retains its initial density structure. Inside, the material is nearly in free-fall and the density asymptotically approaches the form $\rho(R) \propto R^{-1.5}$ and the velocity distribution is $v(R) \propto R^{-0.5}$. This free-fall state implies that the rate at which mass accretes onto the protostar is a constant. The actual value of β is in the range 0.97–1.5, yielding

$$\dot{M} = 1.6 - 2.4 \times 10^{-6} \left(\frac{c_s}{0.19 \text{ km s}^{-1}} \right)^3 \text{M}_\odot \text{ yr}^{-1},\qquad(8.6)$$

for all radii and, in principle, until the envelope is exhausted. This model has been widely explored, partly because it is based on a relatively simple, semi-analytical similarity solution, allowing rotation and magnetic fields to be treated as perturbations.

In comparison, the accretion from perturbed Bonnor-Ebert spheres begins with a sharp rise, peaking at various values but with a maximum achievable accretion rate of $\sim 47\, c_s^3/G$. High rates are not maintained for longer than a few tenths of a free-fall timescale and the accretion rate falls gradually, eventually falling below that of Eq. 8.6.

Another type of collapse is initiated by a strong external trigger rather than a small perturbation. If a static core is triggered by an interstellar shock wave, a high rate of accretion can be produced through the strong shock compression. There is indeed plenty of evidence for triggering, as now listed:

- Cometary clouds take on a wind-blown appearance: a sharp arc-shaped edge or surface and an extended tail. Many contain deeply embedded protostars.

- Molecular clouds face a potential source of a supernova shock or wind from a massive star (e.g. the Orion clouds). Triggered star formation by a distant supernova is also suggested in the Upper Sco association and the ρ Ophiuchus cloud. The star formation in NGC 2265 IRS in the Cone nebula also shows the signs of triggering by a stellar wind from a B2 star.
- Bright-rimmed globules are identified by a curved ionisation front (bright rim) on one side of their surface. They are exposed to strong UV radiation from a nearby OB star. The physical conditions of such clouds are consistent with models of a radiation-driven implosion.
- Groups of young stars sometimes appear to be ordered in space in a sequence of increasing age.
- The Solar system may have also originated from a triggering event. The evidence is derived from studies of meteorites which contain short-lived isotopes. The isotopic abundances in chondrules, in particular short-lived parent radionuclides, could have been obtained from stellar ejecta originating from a supernova and/or an asymptotic giant branch star.

Triggering has been simulated by following the evolution of a Bonnor-Ebert sphere after it has been subjected to an increase in the external pressure. The simulations demonstrate that the main features of the density and velocity field of pre-stellar cores and protostars are reproducible, and that the accretion rate is high in the earlier accretion phase. The predicted velocity field is very different from that of the standard model since the collapse begins from the outside. In many other respects, the triggering model resembles the turbulent cloud model, that we now discuss, since both are based on the impulsive compression of dense regions. For triggering, an equilibrium cloud must first be present.

Other models exist that adopt different initial conditions with various predictions for the density and velocity distributions. Therefore, observational determinations of both of these distributions can help us to discern the physical processes governing the formation of low-mass stars.

8.6 Protostellar Accretion from Turbulent Clouds

Analytical models for turbulence have proven extremely difficult to establish. We therefore rely upon numerical simulations which include as much of the physics as we can manage. Practically all simulations have treated the formation of cores and the evolution within cores as distinct problems.

To date, the 'clumps' found in hydrodynamic simulations of isothermal turbulence possess properties that resemble the observations of cores. While encouraging, the picture is still being assembled. A quasi-equilibrium clump mass spectrum is produced in the simulations with a power-law distribution $dN/d(\log M) \propto M^{-0.5}$. This is consistent with the observed mass distributions of clumps (see Eq. 6.8). The dense, Jeans-unstable gas clumps collapse and form protostellar cores that evolve through competitive accretion and many-body interactions with other cores. In contrast to the clumps, the core mass spectrum is best described by a *log-normal* distribution that peaks approximately at the average Jeans mass of the system. A log-normal is a Gaussian distribution as given by Eq. 3.1 with the variable v replaced by $\log M$. It is the distribution often encountered when many different mechanisms are influencing an outcome. However, the observed distributions of cores differ in that they possess a power-law tail of high mass cores (e.g. cores-rhooph) which is also imprinted in the resulting distribution of stellar masses (§12.7).

The individual collapse of perturbed protostellar cores and their break-up into binaries and multiple systems have been studied extensively. Cores that match the density profiles of both starless and protostellar cores are found. The agreement is certainly remarkable. The density profiles of the majority of cores closely resemble that of the isothermal hydrostatic equilibrium solutions i.e. Bonnor-Ebert spheres. Nevertheless, the cores are far from equilibrium, being dynamically evolving. More evolved cores possess a $1/R^2$ density profile, again consistent with the observations. The next step, yet to be achieved, is to model the velocity profiles of cores, to simulate observed turbulence and infall.

The accretion rates of widely-spread turbulent cores resemble quite closely that of triggered clouds. In a dense cluster environment, the accretion rates are high and highly variable. Since the collapse does not start from rest, very high accretion rates are achievable. High-mass stars tend to form early in the cluster evolution and then maintain high accretion rates. They may form in just a single free-fall time of $\sim 10^5$ years.

In contrast, low-mass stars tend to form later. After a high peak rate, the fall-off can take a wide variety of forms. Tidal interactions with other cores can cause a sudden halt to the accretion. On the other hand, merging of cores and competitive accretion of the available gas produce strong deviations from the smooth evolution of an isolated core. Finally, low-mass cores may find themselves slung out of the cloud by dynamical instability, for example by the gravitational force during a close encounter with another core (see §5.5). Accretion may then terminate abruptly. The conclusion is that the accretion histories of protostellar cores can only be determined statistically.

8.7   Number, Age and Statistics

Class I sources are relatively rare compared to more exposed young stellar objects of Classes II and III found in and around molecular clouds. There are typically ten times less Class I than Class II. Assuming the class system corresponds to an evolutionary sequence, the relative number of sources in each evolutionary phase is proportional to the relative time spent in each phase. This statistical argument then suggests Class I ages in the range $1-5 \times 10^5$ years.

The Class 0 protostars are another factor of 10 less common than the Class I protostars in the ρ Ophiuchus cloud. This can be interpreted as a very abrupt babyhood for a star of $1-5 \times 10^4$ years. To accrete half of a solar mass in this time requires a mass accretion rate exceeding $10^{-5}\,M_\odot\,yr^{-1}$. Alternatively, the scarcity might be interpreted as a peculiarity of the star formation history in this particular region. Specifically, the rate of star formation may have quite sharply decreased about 2×10^5 years ago.

Class 0 sources are found in greater numbers in other locations. In the Taurus region, a region where star formation is more distributed, the duration of the Class 0 phase appears to be longer, $\sim 10^5$ years. Therefore, we should not think of protostars as isolated objects with evolutions determined solely by their initial conditions. More statistical data will help resolve these important issues.

We now also know that most pre-stellar cores must be young and evolve dynamically. Altogether, a good fraction (perhaps around one half) of all cores already harbour young stars. This was first shown in the 1980s by observations with IRAS, the Infrared Astronomical Satellite, which surveyed clouds for infrared sources, presumably a sign of strong accretion from an envelope. This indicates that pre-stellar cores are not long-lived equilibrium gas clouds but must also evolve fast: either collapsing or dispersing through fast dynamical processes rather than slow ambipolar diffusion.

There is further evidence that pre-stellar cores evolve dynamically. Smaller star formation regions contain stars with a narrower range in stellar ages. The duration of star formation corresponds roughly to a few times the cloud crossing times at the turbulent speed. Furthermore, the spread in stellar ages in young clusters is small and comparable to the cluster dynamical time. In these situations, if starless cores could survive longer than 1–3×10^6 years, a range in stellar ages at least equal to this timespan would be predicted. Prime examples are from the Orion Trapezium cluster, the Taurus star formation region, and the NGC 1333 and NGC 6531 clusters.

8.8 Protostellar Evolution

How does the Class evolution occur? Whichever accretion model we choose, we face a problem in evolving a Class 0 source into a Class I source since it is not a transition in which mass is conserved. Instead, most of the core mass may be ejected.

First, it is essential to know how the central object is constructed. As noted above, this determines the radius of the protostar and, hence, the accretion luminosity. There are up to three stages we can tentatively identify for a low mass star. These are as follows:

- An initial growth phase to $0.3\,M_\odot$, possibly with a constant radius.
- A linear growth in protostellar radius with accreted mass controlled by deuterium burning to $1\,M_\odot$
- A small shrinkage as deuterium burning is suppressed, while accretion continues up to $3\,M_\odot$.

A protostar will pass through all these stages only if sufficient mass is available. Higher mass stars will be considered later.

The first stage is rather uncertain. We could take a radius growing from zero to about ~ 2–$3\,R_\odot$ as the mass increases to $\sim 0.3\,M_\odot$. If at constant average density, this would imply $M_* \propto R_*^3$, and, therefore, a rapidly increasing accretion luminosity but no luminosity due to contraction of the protostar itself.

The second stage is controlled by deuterium burning in the protostellar interior. While hydrogen burning does not commence until the star has reached the prime of its life, deuterium ignites when the central temperature reaches about 10^6 K. In a scenario where the protostar accretes at a constant rate of $10^{-5}\,M_\odot\,\mathrm{yr}^{-1}$, this occurs when the protostar has reached a mass of just $0.3\,M_\odot$. The energy input is then sufficient to generate and maintain convection. This means that newly-accreting material is caught up in the convective eddies and rapidly transported to the centre. With constant accretion, we arrive at steady-state fuelling. The luminosity contributed by deuterium is then simply

$$L_D = \delta_D \dot{M} = 12 \left(\frac{\dot{M}}{10^{-5} \, M_\odot \, yr^{-1}} \right) \, L_\odot, \tag{8.7}$$

where we assume a deuterium abundance of $[D/H] = \delta_D = 2 \times 10^{-5}$ and δ_D is the nuclear energy released per unit mass.

The deuterium burning acts as a central thermostat, not allowing the temperature to rise above $\sim 10^6$ K at this stage. In addition, the amount of available nuclear energy is comparable to the gravitational binding energy of the protostar, GM_*/R_*. This implies that the radius of the protostar will increase almost linearly with the mass. A detailed calculation shows that this leads to a $1 \, M_\odot$ protostar with a radius of $5 \, R_\odot$, if sufficient mass is available.

Thereafter, if the accretion continues, the protostar should become radiatively stable. That is, convection is not necessary. Rapid contraction follows until the protostar has accumulated a mass of about $3 \, M_\odot$. If this were to continue, hydrogen burning would soon commence, before the star had finished accreting. Instead, it is thought that deuterium re-ignites, this time in an outer convective shell, causing the protostar to swell up. This later evolution, however, is more relevant to the following chapter.

We can now estimate how the accretion luminosity evolves during a linear $M - R$ phase. Clearly, the luminosity is then directly proportional to the mass accretion rate,

$$L_{bol}(t) = 63 \left(\frac{\dot{M}}{10^{-5} \, M_\odot \, yr^{-1}} \right) \left(\frac{M_*}{1 M_\odot} \right) \left(\frac{R_*}{5 R_\odot} \right)^{-1} \, L_\odot \tag{8.8}$$

and it exceeds L_D at all times, allowing us to equate the accretion power to the bolometric luminosity. In summary, the bolometric luminosity evolves quite closely with the mass accretion rate and can reach values of order 10–$100 \, L_\odot$.

It is remarkable that a star like the Sun reaches a much higher luminosity when young than it will obtain through nuclear burning on the main sequence. The total gravitational energy released in the collapse is given roughly by Eq. 4.7

$$-\mathcal{W} = 1.9 \times 10^7 \left(\frac{M}{M_\odot} \right)^2 \left(\frac{R}{R_\odot} \right)^{-1} \, L_\odot yr. \tag{8.9}$$

If this is released within an accretion phase lasting 1.9×10^5 yr, then the average luminosity is 100 times higher than from the present Sun.

Even more remarkable is that protostellar accretion offers us a great but fleeting opportunity to detect lower mass objects. Even brown dwarfs can produce a high luminosity during their formation. This is important because stars on the main sequence generate nuclear power at a rate $L_{nuc} \propto M^4$. This implies that low-mass stars evolve very slowly but are difficult to find. In contrast, according to Eq. 8.8, although there is less of it, gravitational energy is released at a much higher rate. Therefore, the formation stage is the best possibility we have at present to investigate very low-mass stars.

8.9 Protostellar Envelopes

The events described above are obscured. We cannot directly see a protostar's photosphere because the emitted radiation is absorbed by the dust in the surrounding envelope. The warmed dust then radiates at longer wavelengths. This process may be repeated until the radiation reaches a wavelength which the dust no longer intercepts. From the observer's view, the depth that we can penetrate from the outside into the dust determines a dust photospheric size. The dust photospheric radius depends on the amount of dust and the opacity, which itself depends on the wavelength.

Between the protostar's photosphere and the dust envelope lies an *opacity gap* across which the radiation propagates unhindered. The dust in this region has been evaporated provided the temperature exceeds $\sim 1{,}500$ K. The region is also expected to become progressively vacated as more material falls onto the accretion disk and less heads directly towards the protostar.

The actual physical outer edge to the envelope may lie far beyond the dust photosphere. We define this edge in terms of the core temperature: when the temperature of the core has dropped to that of the ambient cloud, the core is no longer distinguishable. To determine the total mass in the envelope, which is probably the mass reservoir from which a star can be constructed, we need to take a sufficiently long wavelength so that all the dust emission is able to escape from within the photosphere, in addition to the dust outside. Millimetre wavelength observations may thus determine the total dust mass which has been warmed by the protostar, both inside and outside the dust photosphere. This dust mass can be converted to a gas mass, on applying some quite well-tested correlations between the two.

The simplest approach is to model the envelope with power-law radial distributions for the density and temperature. However, there remain tremendous uncertainties in the derived results.

To proceed further, we need to model the parts we usually can't see. An accretion disk should be hidden within the envelope, mediating the flow from the envelope to the protostar. We define an inner edge to the envelope as where angular momentum provides a centrifugal barrier, turning spherical accretion into disk accretion. The envelope loses mass as the system evolves; first to the protostar, then to the disk and then through dispersal. At some stage the envelope becomes optically thin and the disk and protostar will be exposed.

In order to match the observations, we model the mass infall from the envelope as follows. After a sharp initial rise, a peak accretion rate is reached to reproduce the Class 0 stage. This mass passes through the disk, accumulating onto the protostar's surface. As the envelope mass is reduced, the accretion rate from it also falls. This has the effect of reducing the size of the dust photosphere, which tends to increase the bolometric temperature. It also increases the mass of the star relative to the envelope.

These changes should automatically produce a Class I protostar. However, we find that too much mass remains in the envelope, even though the central protostar has accumulated most of its final mass. To make the evolution work, we need the envelope to lose most of its mass in other ways, not to the protostar. How this is achieved remains a mystery although there are processes which could accomplish the deed: disruptive interactions with the environment and dispersing winds driven by protostellar jets.

8.10   Summary: Observation versus Theory

The rates of mass accretion of Class 0 sources are believed to be higher than those of Class I souces. One important line of circumstantial evidence is derived from the CO outflow activity. As noted above, Class 0 sources have an order of magnitude larger power and momentum flux in their outflows than those of Class I sources. Most ejection models predict that the momentum flux of the outflow is proportional to the accretion rate. This implies that these Class 0 sources experience higher accretion rates.

Secondly, and more directly, radiative transfer calculations of infalling, dusty envelopes surrounding Class 0 protostars yield accretion rates exceeding 10^{-4} $M_\odot$ yr^{-1}. Both modelling of the far-infrared peaks in the SED and molecular line modelling support high mass accretion rates. This rate is much higher than the typical accretion rates of $\sim 3 \times 10^{-6}$ $M_\odot$ yr^{-1} of the Class I sources (in Taurus-Auriga), whose values were obtained by similar radiative transfer calculations.

In terms of initially static models, the high accretion rates of Class 0 sources may be explained by beginning from an appropriate density profile. Before collapse begins, a flat inner region surrounded by a power-law envelope is required. In terms of dynamical models, numerical simulations display abrupt collapse and accretion onto dense filaments and cores.

All the isothermal 'similarity' solutions share a universal evolutionary pattern. At early times, a compression wave (initiated by, e.g., an external disturbance) propagates inward at the sound speed, leaving behind it a $\rho(R) \propto R^{-2}$ density profile. At t = 0, the compression wave reaches the centre and a point mass forms which subsequently grows by accretion. At later times, this wave is reflected into a rarefaction or expansion wave, propagating outward (also at the sound speed) through the infalling gas, and leaving behind it a free-fall zone with a $\rho(R) \propto R^{-1.5}$ density distribution.

As already noted above, a problem for the inside-out collapse scenario is the detection of extended zones of infall in cores which have still to develop a recognisable protostellar nucleus. The observed infall asymmetry is spatially too extended (0.1 pc) to be consistent with this collapse model. A related problem is that collapse should set in well before the $\rho \propto 1/R^2$ profile is achieved. The collapse will, however, be mediated by the magnetic field and angular momentum. So this cannot be the complete picture and a modified version of the above scenario could still prove viable.

Chapter 9

The Young Stars

The entity that emerges from a core is a young star that only vaguely resembles the final version. It is spinning, spotted and oversized. The core and envelope have been shed but a massive disk now dominates the environment. In this chapter, we describe and explore the reasons for the violent behaviour and how it is eventually harnessed.

The vast majority of young stars are T Tauri stars. This is simply because they spend most time in this slowly evolving state. It can be argued that the destiny of T Tauri stars has been pre-determined in the previous collapse and main accretion stages. Therefore, this growing phase represents nothing more than a necessary stage to endure in the progress towards adulthood. However, they prove to be of immense observational significance. As objects revealed in the optical for the first time, they have provided most of our knowledge of Young Stellar Objects (YSOs). Due to their large numbers, however, they also provide us with statistical samples and spatial distributions. Finally, we find that both accretion and outflow have not been entirely eradicated but are still ongoing in a moderate form. This activity thus presents itself in manifestations explorable with our optical telescopes.

As planet-forming environments, the classical T Tauri stars have now taken on extra significance. We can explore how a spinning disk might evolve out of an envelope into a stellar system in which a star is surrounded by planets. The disk geometry and the balance of forces permit a more quantitative approach here than for any other stage of star formation. Our concepts concerning the nature and evolution of proto-planetary disks can now be tested. This has led us to re-think our ideas because observations have revealed unexpected disk properties.

9.1   T Tauri Stars

9.1.1   *Classical T Tauri stars*

In 1945, Alfred Joy classified a group of 11 stars as 'T Tauri variables', named after the prototype discovered in the constellation of Taurus. They displayed erratic optical variability, strong chromospheric lines and could be identified through their strong Hα emission lines. They were found to congregate at the periphery of dark clouds. The nearest were located in the Taurus molecular cloud and the ρ Ophiuchus cloud. Given our present knowledge, they clearly represent an early stage in the evolution of stars of low mass (less than $3\,M_\odot$) that are similar to the Sun.

Stars which resemble the original group are now referred to as Classical T Tauri stars (CTTSs). Their temperatures and masses are similar to the Sun but they are considerably brighter. They have been found to possess many extraordinary properties, in comparison to which the Sun is mild-mannered and mature. Their outstanding properties, discussed in detail below, include the following:

- They spin rapidly, with typical rotation periods of 1–8 days, compared to about a month for the Sun (25.6 days at the equator and about 36 days in the polar regions).
- Large areas of the surface are covered by hot starspots, typically 3–20 % whereas the cool Sun spots usually cover a few tenths of a per cent of the solar surface.
- Variable emission occurs in the X-ray and radio bands (a thousand time stronger than solar-like activity).
- Strong stellar winds and collimated outflows are usually found (10^6 times stronger than the solar wind of $10^{-14}\ M_\odot\ \mathrm{yr}^{-1}$). The outflows are often associated with spectacular jets (see §9.8).
- Their external environments are complex. Classical T Tauri stars are surrounded by gas disks, revealed through strong excess emission in the infrared and sub-millimetre (see also Fig. 9.1).
- Most T Tauri stars are in binary star systems (see §12.2.1).

Since these stars are exposed at visible wavelengths, optical spectroscopy provides strong constraints on our interpretation. Ultraviolet observations with the International Ultraviolet Explorer (IUE) revealed high levels of line emission that are typical of the chromosphere and transition regions of the Sun, in addition to the hydrogen Balmer lines, calcium (H and K lines)

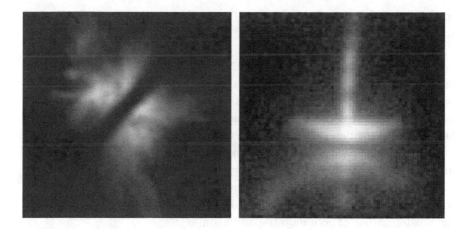

Fig. 9.1 These Hubble Space Telescope images show stars surrounded by dusty disks. You cannot see the star in either photograph, though their light illuminates the surrounding nebula. Left: Image of a star called IRAS 04302 2247. Right: in HH 30, jets emanate from the centre of a dark disk of dust which encircles the young star. The surfaces of the disk can be seen in reflected light. (Credit: NASA, HST, D. Padgett *et al.* 1999, AJ 117, 1490; C. Burrows *et al.* 1996, ApJ 473, 437.)

and various forbidden emission lines. Such lines have traditionally been used to find CTTSs using objective prism techniques. Now, multi-object spectroscopy employing optical fibre technology allows us to survey wide regions quite efficiently.

We now recognise that the hydrogen lines are not actually produced as on the Sun but within an extended dense wind and accretion flow. A strong fast wind is deduced from the Doppler line shifts and widths of the Hα and forbidden lines (e.g. from once-ionised sulphur and neutral oxygen). These lines are often blue-shifted. The absence of red-shifted emission is interpreted as due to occultation of the far side by an opaque disk.

Excess ultraviolet and blue emission often disguise the spectral characteristics of Classical T Tauri stars. The cause is believed to be related to the manner in which material falls from the inner edge of the accretion disk onto the star. The effect is to 'veil' photospheric absorption lines. That is, the absorption lines are less deep than standard stars of the same spectral type. A significant observation is that the veiling is only present if the near-infrared excess is also present.

There are also several T Tauri stars in Taurus whose optical spectra are too heavily veiled to enable detection of the photospheric features needed for spectral classification. These stars are typically referred to as 'continuum' stars.

Ultraviolet emission lines originating from hydrogen molecules have now been identified in a large sample of Classical T Tauri stars. These lines are probably the consequences of resonant fluorescence of molecular hydrogen following excitation from either atomic hydrogen Lyman α photons or pumping from other atomic transitions. The origin of the emission could be either disk or nebular material surrounding the stars. Since H_2 is the most abundant species in the disk, analysis of its physical conditions has important implications for a wide range of topics.

In summary, classical T Tauri stars are viewed as composite systems, comprising of an active stellar photosphere, an accretion zone, an extended circumstellar disk of cool material and atomic jets.

9.1.2 *Weak-line T Tauri stars*

Weak-line T Tauri stars are pre-main sequence stars that have hydrogen lines in emission, but with an $H\alpha$ equivalent width under 10 Å (the equivalent width is the width the line would have on supposing the line emission peaks at the same level as the stellar continuum). In 1978, George Herbig described them as *post T Tauri stars*, having finished their formation and now heading towards the main sequence.

Consequently, large numbers were discovered in X-ray surveys with the Einstein Observatory and ROSAT satellites. These stars were previously not identified as young stars because they also lacked excess infrared and UV emission. Follow-up observations of their radial velocities showed that these serendipitous sources did indeed belong to nearby star formation regions. Locating them on the Hertzsprung-Russell diagram, it became apparent that many of them were not older than the CTTSs. Assigning ages to them according to evolutionary models (Fig. 9.2), it was also clear that these non-disk objects dominate the population of young stars older than 1 Myr.

Due to their unveiling, in the physical sense, the WTTSs were called *naked* T Tauri stars. In general, we refer to the class as weak-line T Tauri star but note that not all WTTSs are stark naked: there are WTTSs which are not NTTSs. Classification is never completely satisfying. Firstly, higher mass G stars have high continuum fluxes which can veil intrinsic hydrogen line emission. Secondly, stars with optically thick disks do not necessarily show strong $H\alpha$ emission, classifying them as WTTSs, yet not considered to be NTTSs.

Among stars which are optically visible, we can extract candidate young stars by searching for the presence of lithium in their atmospheres. Lithium is depleted in stars as they age. It is a fragile element and is destroyed at the temperatures reached at the base of the convection zone. A temperature of $\sim 2.5 \times 10^6$ K is necessary. Convection mixes the material and transports it between the surface and interior. This process, called convective mixing (§9.6), is efficient at reducing the surface lithium abundance. We thus employ lithium absorption lines (specifically a strong Li I 6707 Å absorption line) to help identify and confirm young stars.

9.1.3 *Outbursts: FU Ori objects*

Certain stars in star formation regions have been found to increase in brightness by several orders of magnitude over a period of months to years. These are called FU Orionis eruptive variables, or FU Ori objects. The luminosities can reach a few hundred $L_\odot$. The two best studied members of the class, FU Ori and V1057 Cyg, have both exhibited increases in optical brightness by 5–6 mag on timescales of about a year, followed by gradual fading. The frequency of FU Ori eruptions is still poorly known. The eruptions may occur many times during the T Tauri phase of a star, suggesting mean times of 10^4–10^5 yr between successive outbursts, with decay times of perhaps several hundred years.

Some objects are now accepted as members of the FU Orionis class only because they are spectroscopically very similar to the classical FU Orionis stars even though no eruption has been observed. The classification scheme is now based on other properties of the objects. We use the presence of the overtone bands of CO, observed as deep broad absorption in the infrared, to identify FU Ori disks. The band emission is produced in the hot inner disk through many overlapping lines from the vibrationally excited molecule. In addition, post-eruption optical spectra show peculiar supergiant F and G type stellar spectra, with blue-shifted shell components, and 'P Cygni line profiles' in Hα, taken as evidence for outflow. Although these objects are in an early stage of evolution, it is appropriate to introduce them here since their properties are closely linked to their disks. These objects, such as L1551 IRS5, also produce jets and bipolar molecular outflows. The high luminosities suggest accretion rates exceeding $\sim 2 \times 10^{-6}$ M$_\odot$ yr^{-1}.

A model in which accretion through the circumstellar disk is greatly increased can explain many observed peculiarities of FU Orionis objects, including the broad spectral energy distributions, the variation with wavelength of the spectral type and rotational velocity, and many 'double-peaked' spectral lines in the optical and the near-infrared. Increased mass transfer through the disk may be triggered by a thermal instability or by the close passage of a companion star. Origins for this class of object will be further discussed below (§9.5).

9.2   Class II and Class III Objects

Class II and Class III protostars are characterised by energy distributions which peak in the near-infrared and optical regions of the spectrum. The emission is dominated by that directly from stellar photospheres.

Class II sources are less embedded than Class I sources. Their spectral energy distributions (SEDs) are broader than that corresponding to single blackbodies. The excess emission extends across mid-infrared wavelengths with a power-law infrared spectral index of $-1.5 < \alpha_\lambda < 0$. This emission arises from circumstellar material with temperatures in the range 100 K to 1500 K and is normally associated with a disk system, without a significant envelope. Their bolometric temperatures lie in the range 650 K $< T_{bol} <$ 2,800 K. When observed in the optical, Class II sources typically possess the signatures of CTTSs, and CTTSs generally possess Class II type SEDs.

Class III SEDs correspond to reddened blackbodies. In this case, the emission is associated entirely with a photosphere, with no contribution from an optically thick disk. Without an infrared excess, the radiated energy decreases more steeply longward of 2μm than Class II SEDs, with $\alpha_\lambda < -1.5$. Foreground dust may still provide considerable extinction, which acts to redden the object through the preferential obscuration of the high-frequency end of the intrinsic blackbody. Class III sources produce very little Hα emission, so officially defining them also as weak-line T Tauri stars.

9.3   Location and Number

Young stars may be found distant from their apparent natal cloud or not even associated with any recognisable cloud. Some may well form individually or in isolated systems, as part of a distributed mode of star formation. Others may form in molecular clumps, in a clustered mode.

The age of individual T Tauri stars is sensitive to the chosen model. They can be tentatively placed on the classical Hertzsprung-Russell diagram and located along evolutionary tracks derived from pre-main-sequence models (Fig. 9.2). CTTSs are so found to have typical ages of $1-4 \times 10^6$ yr while WTTSs take a wider range of ages $1-20 \times 10^6$ yr. Most Class III sources are, however, older than 5×10^6 yr, and may well be classified as *post* T Tauri stars.

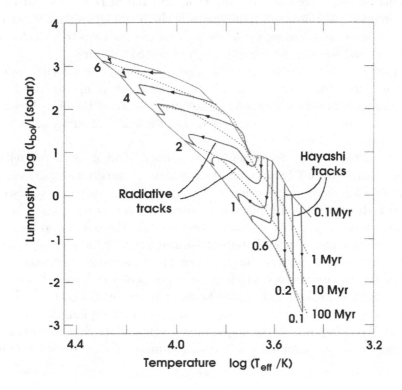

Fig. 9.2 The Hertzsprung-Russell diagram for T Tauri and intermediate mass stars (adapted from work by Palla & Stahler, 1998).

The number of Class III sources is difficult to ascertain since their properties are not so outstanding. As strong X-ray sources, their X-ray emission provides the best means for a census (§9.7). They appear to be at least as populous as Class II sources and to extend further beyond the outskirts of the star formation regions, consistent with their age.

9.4   Accretion

9.4.1   *General characteristics*

There are three stages in the transition from an envelope-dominated to a star-dominated system. Firstly, quite early in the evolution of a forming star, the disk-protostar system is still surrounded by an infalling envelope. The envelope generates the far-infrared SED. At longer wavelengths, the infalling envelope becomes optically thin, and the SED is dominated by disk emission. At this stage, radiation from the infalling envelope is one of the most important heating mechanisms of the disk and has to be included to understand the long wavelength SEDs of embedded stars.

In the second transition phase, the disk and young star are surrounded by a tenuous dusty envelope that scatters stellar radiation onto the disk. This radiation can significantly heat the outer regions of the disk. Finally, in a more evolved phase, it is expected that the material surrounding the disk-star system becomes negligible.

Why a disk? The puzzle was to fit a large amount of circumstellar material around a CTTS without obscuring it. If distributed spherically, the young star would not be visible in the optical. It was therefore suggested that it is distributed in a geometrically thin disk. A disk was, in any case, expected to form as accreting material falls within the centrifugal barrier. The envelope material will land onto the disk and must subsequently accrete from there onto the central object. Direct proof eventually came through the Hubble Space Telescope which imaged spectacular disk-like silhouettes on a background nebula illuminated by stars in the Orion cloud.

Many questions are only half answered. Can a disk explain the infrared excess? What happens to the angular momentum of the disk to allow the material to accrete onto the star? Does the material crash or settle onto the star?

9.4.2   *Accretion disk specifics*

We first focus on what type of disk can reproduce the observed infrared excess. Specifically, we have to match the power-law wavelength dependence of the SED. Physically, we hope to learn about the disk physics although we actually only directly model the temperature structure.

The basic model is that of a disk which is geometrically thin but optically thick. The model disk is dissected into a series of annuli, the surface area of an annulus of radial width ΔR being $2\pi R \Delta R$. The spectrum from each annulus corresponds to a blackbody with a specific temperature. We take a radial temperature distribution $T(R)$. The emergent spectrum is then the superimposed contributions of the annuli. The temperature distribution depends on the heat released from within the disk. If we take a distribution of the form $T(R) \propto R^{-n}$, so that positive index n implies that the inner disk is hotter than the outer disk, then the blackbody luminosity from a disk annulus is

$$L(R)\,\Delta R = 2\pi R \Delta R\ \sigma\ T^4. \tag{9.1}$$

The temperature difference across an annulus is $\Delta T / \Delta R \propto R^{-n-1}$. Substituting for R:

$$L(T) \propto T^{3-\frac{2}{n}} \Delta T. \tag{9.2}$$

Employing Wien's law, Eq. 6.2, $T \propto \nu$, the predicted SED slope possesses a power law slope as defined by Eq. 8.1 of

$$\alpha_{\lambda,disk} = \frac{2}{n} - 4. \tag{9.3}$$

Therefore, the predicted SED is indeed a power law directly related to the temperature distribution across the disk. For Class II sources, the most common values observed lie in the range $\alpha_\lambda \sim [-0.4, -1.0]$, which implies a range of $n \sim [0.56, 0.67]$. The question now is what model for the disk produces such a temperature structure?

First, we consider the standard theory of an 'active' accretion disk. The properties of a standard disk, which channels material from an envelope or mass-losing companion onto the star, are well known. We ignore the pressure in the disk, whether thermal or turbulent, and so balance the centrifugal force with gravity at radius R, given an angular rotation Ω and a dominant central mass M_*:

$$R\Omega^2 = \frac{GM_*}{R^2}. \tag{9.4}$$

This implies that the disk does not rotate like a solid body. It is a sheared flow with rotation rate $\Omega = (GM_*/R^3)^{1/2}$.

We can now understand why the material in the disk will fall inwards despite the necessary conservation of total angular momentum. Viscous forces are created in the disk due to the shear, as described in §5.3. Viscosity leads to dissipation of flow energy into thermal energy i.e. the disk is heated. It also leads to outward transport of angular momentum and a redistribution of energy: the specific angular momentum, $\Omega R^2 \propto R^{1/2}$, decreases with radius instead of being conserved.

In a steady-state, the rate at which mass flows inwards is just the rate at which we supply the disk with new material. The dissipation through viscosity should adjust to meet the demands: it can redistribute angular momentum and energy but not alter the totals. This property of the viscous force was also found in the discussion of shock fronts (see §5.5). If the mass supply is high, then the disk will be massive so that molecular friction can cope. If the mass supply is too high, however, then molecular viscosity is ineffective and some form of anomalous viscosity must take over. Turbulent viscosity is generated by MHD instabilities (see §9.5).

Geometrically, the disk may be thin but it has a thickness which is not constant. We define the thickness by the pressure scale height $H(R)$, a function of the disk radius R. If the spinning disk is assumed to be in vertical hydrostatic equilibrium, then a vertical pressure gradient pushing up is balanced by the small component of radial gravity which pushes down. This equilibrium implies

$$H/R = c_s/v_d \qquad (9.5)$$

where v_d is the Keplerian rotation speed ΩR.

Finally, if the disk mass is so high that self-gravity becomes important, then gravitational instability prevails and tidal torques transport the angular momentum outwards and the mass inwards. Gravitational instability also leads to fragmentation and the formation of bound objects: brown dwarfs or planets. For a disk whose potential is dominated by the protostar and whose rotation curve is therefore approximately Keplerian, instability sets in when the disk surface density Σ is sufficiently large. Analysis yields the condition for instability:

$$\Sigma > \frac{c_s \Omega}{\pi G Q_c}, \qquad (9.6)$$

where Q_c is called the critical value of the so-called Toomre parameter (of order unity). Putting the disk mass $M_d \sim \pi R_d^2 \Sigma$ and using Eq. 9.5, we obtain the instability condition

$$M_d > \frac{H}{R} M_*. \qquad (9.7)$$

Therefore, a thin disk which contains a significant mass fraction will be unstable. For the outer parts of YSO disks $H/R \sim 0.1$, and thus a massive disk is required for instability.

Remarkably, we already know enough to predict the temperature distribution. The energy for the local dissipation ultimately derives from the gravitational potential. The heating extracted directly from the infall is at the rate

$$\dot{E} = \dot{M} \frac{\Delta R}{R} \frac{GM}{R}.$$ (9.8)

This is mollified by the change in rotational energy and modified by the viscous transport of energy. We also place an inner boundary condition where the dissipation rate falls to zero. It turns out that the energy release maintains the above form at large radii and can be written in more detail as

$$\dot{E} = \frac{3GM\dot{M}}{2R^2}\left[1 - \left(\frac{R_*}{R}^{1/2}\right)\right] \Delta R.$$ (9.9)

Integration over the entire disk yields a total dissipation of $GM\dot{M}/(2R_*)$, exactly one half of the available potential gravitational power. The other half remains in the form of the inner spinning disk, from where it must be somehow extracted. Hence, the disk is only half the potential story.

The effective temperature of the optically thick disk is then given by the formula for a blackbody, Eq. 6.4, taking into account that the disk has two surfaces, $\dot{E} = 4\pi R \Delta R \, \sigma_{bb} \, T_{disk}^4$:

$$T_{disk}^4 = \frac{3GM\dot{M}}{8\pi\sigma_{bb}} \frac{1}{R^3}\left[1 - \left(\frac{R_*}{R}^{1/2}\right)\right].$$ (9.10)

In other words, $T_{disk} \propto R^{-3/4}$ over the extended disk.

This is not quite what we were hoping for: a standard disk is predicted to have a radial temperature distribution with $n = 0.75$ whereas the Class II SEDS generally imply a shallower fall-off in temperature, $n \sim [0.56, 0.67]$.

A second problem is also raised by the other half of the story: while half the disk accretion power may appear as the infrared excess, the other half must be lost in a small boundary layer between the disk and the star. This high power from a small area would produce ultraviolet excess emission which should be as strong as the infrared excess. Indeed, an ultraviolet excess is observed. However, the excess is far less than the infrared excess for many well-studied systems. This suggests that the mass accretion rate onto the boundary layer is actually lower than that through the disk, which would cause material to pile up in the inner disk. Such a situation cannot last long. This leads us to suspect that we have not taken everything into account.

The solution appears to lie in the fact that we assumed the disk to be isolated. It is, in fact, exposed to the central star. A major heating agent in many CTTSs will be irradiation: a passive disk is heated on processing of the incident radiation. For a flat disk, however, this turns out to yield the same temperature distribution as predicted from the standard model. The reason for this is that the stellar luminosity $L_* = 4\pi R_*^2 \sigma_{bb} T_*^4$ is diluted by $(R_*/R)^2$ due to the usual spherical expansion plus a further R_*/R due to the grazing incidence of the radiation. In detail, the effective disk temperature due to the irradiation is given by

$$T_i^4 = \frac{2T_*^4 R_*^3}{3\pi} \frac{1}{R^3}. \tag{9.11}$$

Therefore, on comparing Eqs. 9.10 and 9.11, the irradiation dominates what we measure from the entire disk provided the accretion rate is less than

$$\dot{M}_{crit} = 2 \times 10^{-8} \left(\frac{T_*}{4000\,\text{K}}\right)^4 \left(\frac{R_*}{2\text{R}_\odot}\right)^3 \left(\frac{M_*}{0.5\text{M}_\odot}\right)^{-1} \text{M}_\odot\,\text{yr}^{-1}. \tag{9.12}$$

For a flat disk, however, at most 25% of the stellar radiation is directed sufficiently downwards to intercept the disk (none from the pole and 50% from the equator). Yet, for many CTTSs the infrared excess accounts for as much as 50% of the total stellar radiation (e.g. for T Tau itself).

What accretion rate is predicted for the disk? An anomalous viscosity which generates the turbulent transport of angular momentum, might scale in proportion to the disk scale height and sound speed according to §5.3. We thus take the viscosity to be $\alpha_{ss}c_s H$, a prescription for a standard disk, first invoked by Shakura & Sunyaev in 1972. Then, using the above analysis, the accretion rate can be written

$$\dot{M}_{disk} = 3\alpha_{ss}\frac{c_s^3}{G}\left(\frac{M_d}{0.01M_*}\right)\left(\frac{0.1R_d}{H}\right). \tag{9.13}$$

Written in this form, one can see the type of disk necessary to convey the envelope mass onto the protostar, given the supply from the envelope given by Eq. 8.5.

If the disk is flared, so that the disk 'photosphere' curves away from the disk midplane, the cool outer regions receive more stellar radiation and so are gently warmed, exactly as required to produce a shallower temperature gradient. Moreover, the extra heating contributes to increasing the height of the disk and so to increasing the exposure. As we have discussed above, a flared disk is actually predicted. Given the vertical scale height $H \sim c_s/\Omega$ from Eq. 9.5 and substituting for $c_s \propto T_{disk}^{1/2}$ and Ω, yields $H/R \propto R^{1/8}$ – a very gradually flaring disk exposed to the protostar.

9.4.3 *The star-disk connection*

It was long thought that the disk continued all the way down into the star and disk material joined the star via a hot boundary layer where half the accretion energy would be released. There is now a large and growing body of evidence which suggests this does not happen. Instead, the current paradigm asserts that the stellar magnetic field is strong enough to disrupt the inner disk. The disk is truncated and the material is then diverted along the magnetic field, forced to accrete nearly radially onto the star, as shown in Fig 10.4.

A key prediction of the magnetospheric infall scenario is the formation of a shock wave on the stellar surface. Material reaches the surface at supersonic velocities, approaching the surface at a free-fall velocity of several hundred kilometres per second. Therefore, close to the surface the gas produces a strong standing shock wave, in which the kinetic energy of the infalling stream is transformed into random thermal motions. It is thought that radiation from the shock, half reprocessed by the stellar surface, is responsible for the UV and optical continuum excess which veil the absorption lines in CTTSs.

Magnetospheric accretion also implies that there is an inner hole in the disk. The disk is truncated close to or inside the co-rotation radius, which is typically at radii 3–5 R_*. There will then be no contribution to the infrared spectrum corresponding to emission from gas at temperatures exceeding about 1000 K. Such 'inner holes' would thus correspond to a sharp drop in emission in the near infrared. This effect is observed although other contributions and geometric effects can hide it.

The peculiar spectral characteristics of classical T Tauri stars are now usually interpreted in terms of magnetospheric accretion. The densities in the accretion column, 10^{10}–10^{14} cm^{-3}, are typical of the solar chromosphere and transition region. The infalling material is detectable as a redshifted absorption feature in the wings of prominent broad emission lines. The magnetic field, and hence the accretion streams, will rotate with the star, leading to variability of the strength and velocity of the absorption component. Observations of this shifting absorption can be used to constrain the geometry of the inner regions of the system, and hence the geometry of the magnetic field.

The accreting column can be divided into three regions: the shock and post-shock regions, the photosphere below the shock, and the pre-shock region above it. The shocked gas is heated to a high temperature, of order 10^6 K, which then decreases monotonically as the gas settles onto the star. It emits mostly soft X-rays and UV. Half of this radiation is sent to the photosphere below and half to the pre-shock region above. In turn, a fraction of the radiation from the pre-shock gas is emitted back toward the star, adding to the radiation from the post-shock gas to heat the photosphere. The flux of radiation incident on the photosphere heats the gas to temperatures higher than the relatively quiescent surrounding stellar photosphere, producing a hot starspot.

The origin of the veiling in the accretion shock column, rather than in a boundary layer, has crucial energetic consequences. In the immediate vicinity of the boundary layer, disk material is rotating at nearly Keplerian velocity at the stellar radius; to enter the star, the material has to slow down and emit this kinetic energy, which amounts to nearly half of the accretion luminosity while the disk emission accounts for the other half. In contrast, in magnetospheric accretion, material falls from several stellar radii, so that the energy released is almost the entire accretion luminosity. Therefore, the intrinsic disk emission is only a small fraction of the accretion luminosity.

9.5   Class and Disk Evolution

As it is unfolding in this chapter, a young star undergoes a complex development. It involves the evolution of several components within a single system, and the transfer of mass and radiation between the components.

The mass accretion rates for CTTSs derived from veiling measurements lie in the wide range of 4×10^{-9} to 10^{-7} $M_\odot$ yr^{-1}, much lower than the rates derived for Class 0 and Class I protostars. Modelling the accretion column and calibrating emission lines in the near-infrared, has allowed us to also tentatively assign accretion rates onto even optically-obscured protostars. This method indicates that many Class I protostars appear to be slowly accreting from their disks even though the envelope is supplying material at a high rate onto the outer disk (see §8.10). In Taurus, Class I disk accretion rates occupy a wide range around $\sim 10^{-7}$ $M_\odot$ yr^{-1} whereas, on average, the infall from the envelopes proceeds at the rate

$\sim 3 \times 10^{-6}$ M$_\odot$ yr^{-1}. This anomaly is termed the luminosity problem for Class I sources.

One suggested resolution is that mass indeed accumulates in the outer disk. When sufficient mass has accumulated, the disk becomes gravitationally unstable, with subsequent strong and abrupt accretion until the excess disk mass has emptied. This would account for the outbursts which identify FU Ori objects (see §9.1.3). A more likely cause for the outbursts is that of thermal instability in the inner disk which is expected when the mean accretion rate is quite high – above 10^{-6} M$_\odot$ yr^{-1}. An objection to such scenarios is that very few FU Ori disks have been found so far although protagonists will claim that they are still deeply obscured. If common, then it is no longer possible to locate young stars unequivocally on evolutionary tracks.

The infall phases correspond to the Class 0 and Class I protostars. The mean mass left to accrete during the Class II phase is just ~ 0.01 M$_\odot$ Nevertheless, the early accretion is probably episodic with short periods of high accretion.

Despite the outbursts, the long-term accretion rate appears to decrease with age from Class I to Class II sources. Using statistics, we derive a relationship of the form

$$\frac{\dot{M}}{10^{-6}\ \text{M}_\odot\ \text{yr}^{-1}} \sim \left(\frac{t_{age}}{10^5\ \text{yr}}\right)^{-1.5}, \tag{9.14}$$

according to the present data, which is probably strewn with errors. Nevertheless, if we assume that exactly this mass is fully supplied by the disk, we can compare the predicted disk mass with that derived directly from submillimetre observations of the dusty disks. The two are in reasonable agreement (for an appropriate dust opacity in the disk, needed to convert dust mass into gas mass).

Can we explain the above power-law evolution of mass accretion? For the disk to evolve, angular momentum must be transported. Little can be transferred to the star itself without it reaching break-up speeds. We have essentially three choices, as follows:

- The disk gas coagulates into dense bodies which then sweep clear the remaining gas.
- A disk wind is driven by centrifugal forces, removing a high fraction of angular momentum.
- Disk viscosity transfers angular momentum outwards.

On the largest scales, the third choice is favoured. In this case, as the system evolves, the disk grows in size in order to incorporate more angular momentum into less material. We note from Eq. 3.17 that the angular momentum is $J_d \propto M_d \Omega_d R_d^2$ in a disk of evolving mass M_d and radius R_d. From Eq. 9.4, we can substitute $\Omega_d \propto R^{-3/2}$ to yield $J_d \propto M_d R_d^{1/2}$. Thus, most of the angular momentum lies on the outskirts of the disk, and, as the disk loses mass, it expands at the rate $R_d \propto M_d^{-2}$, conserving angular momentum.

The evolutionary timescale t_{visc} depends on the viscous transport. As discussed in §5.3, this is a slow diffusion process with

$$t_{visc} = R^2/\nu. \tag{9.15}$$

for a rotating disk. To proceed from here we now need a viscosity prescription. Given the shear, we expect that turbulent viscosity will dominate. This takes the form of Eq. 5.5: $\nu_e = \lambda \times v$. We substitute turbulent length and velocity scales appropriate for the disk: the scale height and sound speed and take an efficiency factor α_{ss}, to yield

$$\nu_e = \alpha_{ss} c_s H \propto c_s^2(R)/\Omega(R) \propto T R^{3/2} \tag{9.16}$$

to describe a standard or 'alpha' disk model. Substituting $T \propto R^{-0.6}$ (see above), then $\nu_e \propto R^{0.9}$. The timescale thus predicted from disks we see, interpreted as standard disks, is $t_{visc} \propto R^{1.1}$. This implies that as a disk expands, the evolutionary time increases and, from angular momentum conservation, the disk mass $M_d \propto t^{-1/2.2}$. This then yields a mass transfer rate of $\dot{M}_d \propto M_d/t \propto t^{-1.46}$. This is remarkably close to the relationship we derived between accretion rates and estimated ages of Class 1 and CTTSs (Eq. 9.14), suggesting that we might be reaching a fully consistent basis for understanding disk behaviour.

The physics behind the viscosity has attracted much attention. We have not been specific about the cause of the disk turbulence. We require a hydrodynamic or magnetohydrodynamic instability to disturb the sheared disk and tap its energy. Values for the disk parameter $\alpha_{ss} \sim 0.01–0.1$ are consistent with observations. One instability mechanism is the magneto-rotational instability (MRI), also called the Balbus-Hawley instability after the discoverers. This is based on the interchange of the positions of magnetic field lines which thread vertically through the disk, leading to a loss of equilibrium. If the disk mass is sufficiently large (see Eq. 9.7), then self-gravity is relevant. Self-gravity will generate density structure in the form of spiral arms. Gravitational torques may then simulate a viscosity. The viscous heating may increase the sound speed and so hold the disk mass to the critical value $M_d \sim Mc_s/R_d$.

9.6 Interiors

Theory predicts that a low mass star does not progress from a dim and cold beginning to become a luminous and warm star in a steady increasing fashion. On the Hertzsprung-Russell (H-R) diagram, the evolutionary path is seen to perform a detour along what is termed a Hayashi track (Fig. 9.2). Here we explore the reason for the detour.

When the envelope has been shed, the young star can be observed in the visible. Since it is highly opaque, the gravitational energy released through contraction cannot easily escape. It possesses a photosphere, interior to which photons are trapped. For this reason, it is a quasi-static object in which the internal pressure is very close to supporting the object against gravity. This state is called hydrostatic equilibrium.

As the object slowly contracts, gravitational energy is released. Half of this energy remains as internal energy, helping to maintain the pressure support (a requirement of the virial theorem (see Eq. 4.4)). The other half escapes from the photosphere as radiation.

A major difference between young stars is brought about by the manner in which energy is transferred to the surface. It was originally thought that the internal structure is simply determined by the means by which the photons would leak out. This is the process of radiative diffusion: the journey of a photon is a random-walk, being repeatedly emitted and absorbed by the atoms.

T Tauri stars behave differently. It was shown by Hayashi in 1966 that their interiors begin by being in a state of convection. This means that if a packet of gas is displaced radially its displacement will not be suppressed but will grow. For example, a packet of gas rising from deep in the star into a lower pressure region nearer the surface will expand. If it expands sufficiently, it will remain buoyant in its new surroundings and continue to rise. Convection thus redistributes the energy by circulating the gas.

Convection becomes a very effective energy transport mechanism in young stars as soon as there is a steep temperature gradient to drive it. This occurs in T Tauri stars because radiative diffusion cannot cope with the prodigious internal energy release as the young star becomes opaque.

The internal temperature rises, setting up convection currents. During this phase, the energy release no longer depends on the efficiency of internal transport but just on how quickly the surface layer can liberate it. The surface opacity is, however, very sensitive to the temperature, providing a positive feedback which tends to stop the temperature from changing. The result is that, as the star contracts, the surface temperature stays approximately constant while the luminosity falls as $L \propto R_*^2$ (i.e. according to the blackbody formula, Eq. 6.4). This yields the near vertical evolution on the H-R diagram shown in Fig. 9.2.

As the luminosity falls, the opacity and temperature gradients fall in the inner regions. Consequently, the core of the star becomes radiative. The size of the radiative core grows, leaving a shrinking outer convective zone. In this radiative phase, the internal radiative diffusion, rather than the surface opacity, controls the energy loss. The contraction is now slow but, because the star gradually contracts, the internal energy must rise (according to the virial theorem) and so the central temperature rises also. The result is that the luminosity increases. Both the increase in luminosity and the contraction ensure that the temperature rises relatively fast. The end result is a quite horizontal evolution on the H-R diagram, along a 'radiative track'.

There will probably always be considerable uncertainty surrounding pre-main sequence evolution. Besides well-known uncertainties in the opacity and in the means used to model convection (mixing-length theory), the accretion of residual disk gas is important. Nevertheless, the theoretical tracks provide masses and ages for individual objects which are roughly consistent with other estimates. We estimate that a solar-type star will spend 9 Myr descending the Hayashi track and then 40 Myr crossing to the main sequence on a radiative track. In contrast, stars with mass above about 2.5 $M_\odot$ possess no convective track while those with a mass exceeding perhaps 6 $M_\odot$ cannot be considered to experience such pre-main sequence phases. Instead, they probably finish their main sequence life before a contemporary low-mass star has contemplated it.

9.7   Giant Flares, Starspots and Rotation

The surfaces of T Tauri stars are extremely active. The activity provides further clues as to how the young stars evolve. However, many results do not as yet fit neatly into any prevailing scenario.

T Tauri stars are strong X-ray sources. The X-ray emission is roughly correlated with the bolometric luminosity by $L_X \sim 10^{-5}$–10^{-3} L_{bol}. In comparison, $L_X/L_{bol} \sim 10^{-6}$ for the Sun and $L_X/L_{bol} \sim 10^{-7}$ for massive stars. The radiation is continuous in wavelength, generated from the thermal Bremsstrahlung mechanism introduced in §2.3.1, except in this case from a very hot plasma (ionised gas) with a characteristic temperature of 1 keV, or 10^7 K.

The origin of the X-rays lies in strong magnetic activity. The interior convection and rotation lead to differential rotation within the star. This generates magnetic flux through a dynamo mechanism. The flux is released sporadically with magnetic field-line reconnection significant, disrupting and heating the atmosphere. Considerable evidence points to this solar-like paradigm. In particular, variations in the X-ray luminosity by factors of a few over months and factors of 10 over days have been recorded. The variations are flares: an abrupt rise followed by an exponential decline. The temperature falls as the flux falls, consistent with a cooling plasma. The flares are gigantic, sometimes stretching out over several stellar radii.

In general, WTTSs are stronger than CTTSs in the X-rays. This is at least partly because the CTTSs are fairly heavily absorbed, hence not found so easily detected in X-ray surveys. However, there are probably intrinsic differences. In addition, protostars and brown dwarfs also appear to be X-ray sources, especially Class I protostars. Moreover, peak X-ray luminosities can be extremely high. The full meaning of all these results is yet to be assessed.

Painstaking monitoring campaigns have revealed that most T Tauri stars display periodic luminosity variations from a fraction of a day up to about 10 days. The variations are often approximately sinusoidal, consistent with large spots on the surface of rotating stars. The area occupied by the spots can be between 2–30% of the surface area. In WTTSs, spectroscopic fluxes indicate that the spots are cooler regions and an explanation in terms of convection and magnetic activity is appropriate. The spot interpretation is confirmed through high resolution spectroscopy of individual absorption lines in the photospheres. The width of these lines corresponds to the Doppler shifts caused by the rotation. Clearly, for those stars observed along the rotation axis, we will detect neither the spot-generated variations nor the rotation speed. For the majority, however, we have measurements of the rotation period, P, and the radial component of the rotation speed, $v \sin i$ (i is traditionally used to denote the orientation of the rotation axis to the observer). The radii of the stars are also given from the blackbody component of the luminosity and temperature. We thus obtain comparable sets of data for the speed of rotation which confirms the periodic variations to be caused by surface blemishes. .

The spots on classical T Tauri stars are hotter than the surface. This implies an alternative origin linked with high-speed infall from the accretion disk, as discussed in §9.4.3. The spots are thus directly connected to the spinning disk.

This brings us to a long-standing 'conundrum': where does the angular momentum go during the evolution of a young star? Having contracted from their natal cores, young stars should spin close to their break-up speeds (i.e. the rotation speed should approach the gravitational escape speed). In fact, their typical rotation speeds of 10–25 km s^{-1} are only about 10% of their break-up values. Subsequently, the young stars should spin up considerably as they accrete from the rotating disk and contract along Hayashi tracks. In fact, they are still slow rotators when they eventually arrive at the main sequence.

There is some evidence for a spin up. In the Taurus region, the distribution of rotation periods is bi-modal with the CTTS possessing periods 5–8 days and the WTTSs periods 1–4 days. The gap between these ranges is the subject of much speculation. For T Tauri stars in the Orion Nebula Cluster the bi-modal distribution is again present but only for stars with masses exceeding $\sim 0.25\,M_\odot$. Furthermore, there is a subset of young stars rotating with a high fraction of the break-up speed.

How can we answer the conundrum? In very young systems, it is thought that the angular momentum of the star is strongly regulated by the circumstellar disk. The stellar magnetic field threads the disk, truncating it inside the co-rotation radius (i.e. where the disk rotation period is less than the stellar rotation period, perhaps at about $5\,R_*$). The accretion then continues along the field lines, producing hot surface zones at high latitudes. Simultaneously, magnetic torques transfer angular momentum from the star into the disk. Such a disk-locking or disk-braking mechanism seems essential to explain the low angular momenta of CTTSs. Later, when the evolution has slowed, magnetic wind braking can be more effective.

9.8   Summary

There are several defining moments in the life of a young star of mass under $\sim 2.5\,M_\odot$. One special transition occurs when it loses its opaque accretion disk. The lifeline to the maternal cloud is irrevocably broken and it stands alone, subject to its changing interior structure. Surface blemishes and giant flares testify to internal upheavals in an exaggerated version of the adult star that it develops towards.

It had already experienced a more dramatic transition for us when it lost its envelope, becoming exposed to visible light. The accretion disk and photosphere then became open to a deep inspection of their atomic compositions. For the star itself, the transition was less critical since the circumstellar disk remained in place, providing an outlet to release its rotational energy. Throughout, however, the disk needs to be churned to feed the star: some form of turbulence separates the spin from the gas.

There is a remarkable twist to the story. Another activity has been ongoing and evolving while all the transitions since the birth have been taking place. This activity – the bipolar outflow – has radically altered many of our interpretations: we have underestimated the ability of a protostar to influence its own environment.

Chapter 10

Jets and Outflows

The ejection of prodigious amounts of material during birth still puzzles us. It is paradoxical that just when we expect to observe infall, the signatures of outflow prevail. Ultimately, we want to know why. Are we witnessing simply the disposal of waste material – the placenta being cast aside? Or are outflows essential, a means of removing some ingredient which allows the rest to flow inwards? There are reasons to believe it is the latter, in which case outflows hold the key to the formation of individual stars.

Outflows may also determine the destiny of entire clusters of stars. They remove gravitational energy from a small scale and feed it back into the large-scale cloud. Done at high speed, the cloud is disrupted and dispersed. The outflow feedback thus influences the star formation efficiency. The young cluster stars are then subject to less gravitational pull and, consequently, also disperse into looser associations and, eventually, the general field.

All protostars and all young stars with accretion disks appear to drive powerful outflows of some kind. Outflows are driven by accretion. If we can understand how the outflow evolves as the accretion proceeds, we can perhaps understand how the star evolves. First, we need to investigate from where this material originates and the launching mechanism in operation. Progress has so far proven slow because the structure of each outflow is different: the determining factors are as much environmental as genetic. A variety of spectacular displays have been uncovered over the last twenty years and the common factors and causes are still not established.

Outflows were at first thought to be of no great consequence to star formation and of restricted influence on their environment. They were thought to be merely outbursts lasting ten thousand years during the million-year evolution. This changed with the discovery of gigantic old outflows and the realisation that the early protostellar stages are also short term. Together, these facts suggested that many outflows maintain a record of events over the entire protostellar history as well as providing an account of present activity.

10.1 Classical Bipolar Outflows

In the 1980s, outflows of molecular gas were discovered around many of the youngest stars in star formation regions. Originally, these were mainly associated with high mass stars. The outflows were found to be split into two lobes, one on each side of the young stellar object. Most significantly, the material on one side was found to be predominantly blue shifted and on the opposite side red shifted. This, in itself, does not require the gas to be part of an outflow but other factors leave no other option open. For example, the radial speeds implied by the line shifts are often of the order of tens of km s^{-1} on scales larger than 0.1 pc. Such speeds can only be realistically produced from quite near the protostar. Other characteristics, in particular proper motions and jets, have confirmed that these are really outflows.

These bipolar outflows had remained hidden to us until we had developed the technology to observe cold molecular gas. They are best observed in emission lines from the CO molecule at millimetre wavelengths. Employing different transitions and isotopes, we gain information about the temperature, opacity and density, as discussed in §2.3.2. Now, we employ a wide range of molecular tracers and techniques.

Such analyses may give clues to the entire history of the protostar, provided we learn how to decipher what we observe. We can search for signs that the outflow was stronger or more collimated when the driving protostar was younger. If an outflow lobe which now has a linear size of 1 pc has grown at the speed of 10 km s^{-1}, then it must be 10^5 years old, comparable to the age of a Class I protostar. In general, outflow sizes range from under 0.1 pc to several parsecs. Furthermore, outflow speeds range from a few km s^{-1} up to about 100 km s^{-1}. Therefore, kinematic ages take on a wide range of values. It should be clear, however, that the kinematic age may not represent the true age.

Also, the total mass in the outflow covers a wide range, from $10^{-2}\,\mathrm{M_\odot}$ to $10^3\,\mathrm{M_\odot}$. The mass set in motion, especially in the youngest and most powerful outflows, far exceeds the mass accreted onto the protostar. Therefore, while processes near the protostar drive a bipolar outflow, it does not provide the material. In other words, we observe material which has been swept-up or entrained.

Even before we could map the features, it was long known that only a small fraction of a lobe produces the molecular emission. The structure within the bipolar lobes can now be resolved, e.g. in HH 211 (Fig. 10.1). We detect thin shells, further indicating that material has been swept up and compressed. Shell or limb structure implies that there is a cavity from which molecular material has been largely evacuated. In other cases, the emission is from numerous clumps spread more evenly. These clumps could be a consequence of fragmentation of the shell.

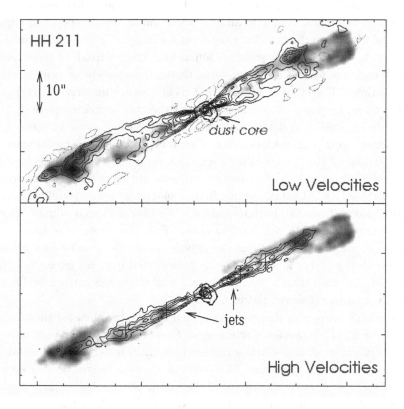

Fig. 10.1   The HH 211 outflow in CO rotational emission (contours) split into two radial velocity components (top panel displays only speeds within 8 km s−1 of the maternal cloud). The H$_2$ near-infrared emission (greyscale) shows the warm shocked gas. (Credit: F. Gueth & S. Guilloteau, Astron. Astrophys. 343, 571-584 (1999).)

There are possibly two distinct types of bipolar outflow. The 'classical' outflows, which dominated those found in the 1980s, are weakly collimated: wide relaxed-looking structures. The collimation factor, q_{coll}, defined as the ratio of the lengths of the major to minor axes, is used to distinguish the type of outflow. Classical outflows possess values $q_{coll} < 4$. Of course, we view just the projection onto the plane of the sky and projection could cause a highly collimated outflow, orientated close to the line of sight, to appear very wide. However, there is no equivalent sample of highly collimated outflows to suggest that the low collimation is purely an effect of projection. Classical bipolar outflows have quite low kinematic ages, in the range 2–5 $\times 10^4$ yr. In contrast, outflow statistics suggest an outflow duration of 2 $\times 10^5$ yr.

Quadrupolar outflows possess two red and two blue shifted lobes. Their arrangement is consistent with an interpretation as two independent super-imposed bipolar outflows. There are several examples, including Cepheus E and L 273. Multiple young stars are found near the centroids of these lobes. In fact multiple sources are often found through deep radio or near-infrared observations. Therefore, it appears that if each source undergoes an outflow phase, then the phase cannot last for the total star formation period.

A major goal is to determine how the momentum, however ejected, is transferred to the ambient medium. Combined spatial and radial velocity distributions of the gas contain information on how the gas is accelerated. The radial velocity data of numerous outflows display a pattern referred to as a 'Hubble Law'. That is, the maximum velocity increases roughly linearly with distance from the outflow source. We also find that higher speeds are observed closer to the outflow axis. Two interpretations have been advanced: we may be observing the expansion of one or more bow-shaped shells with the overall lobe shape being preserved during the expansion. Or, material is ejected with a range of speeds and directions during outbursts, directly into the observed pattern.

The high thrusts of the outflows pose a serious problem for their origin (see §10.8.2). The transfer of momentum from a stellar wind would have to be very efficient. A high efficiency of momentum transfer could be obtained if the flow were energy driven. This requires a means to transfer the energy into the ambient medium instead of the momentum. This is not easy to accomplish except in a spherical expansion within which the accelerated gas remains hot. The observed flows, however, do not possess large inflated bubbles or predicted high lateral motions. Instead, bipolar outflows often display strong forward-directed motion.

10.2 High-collimation Bipolar Outflows

During the 1990s, a new class of bipolar outflow was revealed. These outflows are distinguished by their high collimation with q_{coll} occasionally exceeding 20. The measured collimation depends on the observed radial velocity: the highest collimation is found when we restrict the data to just the highest velocity gas.

There is a continuity between the classical and the high-collimation outflows. The many common features include shell structures, Hubble laws, and kinematic ages. In addition, the new outflows possess gas moving with extremely high velocities (EHV). What makes the speeds, exceeding 50 km s^{-1}, in some sense extreme is that we would expect molecules to be destroyed during the acceleration process if a shock wave were responsible. Sometimes, the EHV component is contained within compact clumps, hence termed 'bullets'. In the L 1448 outflow from a Class 0 protostar, the bullets are symmetrically placed about the driving protostar.

Of great significance to the star formation story is that the outflows with high collimation are mainly driven by Class 0 protostars whereas the classical outflows are driven mainly by Class I protostars. Furthermore, the highly collimated flows are more energetic than classical outflows when compared to the protostellar luminosity. It is thus apparent that the outflow power is strongly dependent on the accretion rate. Outflows are usually characterised by the momentum flow rate rather than the power since this quantity, the thrust, can usually be more accurately measured. In terms of evolution, there is evidence that outflows become less collimated as the protostar evolves.

Spectroscopic differences are also found. The main diagnostics are those of CO line profiles which are interpreted in terms of the fraction of mass accelerated to high speeds. The high-collimation flows tend to have relatively high mass fractions at high speeds while more evolved outflows, as well as the more massive outflows, tend to contain lower fractions of high speed gas. The obvious interpretation is that the injection of high-speed gas decreases and the reservoirs of slow-moving gas accumulate as an outflow ages. The data are however rather fragmentary and various effects are difficult to separate.

Analysis of the radial velocity also indicates that the gas is in 'forward motion' in the higher collimated outflows. That is, there is very little motion transverse to the axis. This indicates that the sweeping is not through simple snowploughing since a plough would deflect material sideways. Instead, the momentum delivered from the source is efficiently transferred. On the other hand, classical bipolar outflows often display overlapping blue and red lobes and multiple components, consistent with their wide opening angles. Overlapping outflow lobes have also been very commonly associated with X-ray emitting protostars. This suggests the existence of thick obscuring disks with a rotation axis orientated close to the line of sight. In such pole-on configurations, the X-rays escape through the outflow cavity.

The significance of the disk in directing the outflow is now well established. The outflow axes are often directed transverse to circumstellar disks where resolved. HH 211 is an excellent example of this (see Fig. 10.1 & 9.1).

10.3   Molecular Jets

Jets are now believed to be present in all young stellar objects in which infall is taking place. The word 'jets' is rather a loose term for all slender structures which appear to emanate from close to a young star and have the appearance of being high speed directed flows.

Jets from Class 0 protostars are found to be extremely powerful, dense and molecular. The structures are usually observed in the near-infrared through their emission in lines of molecular hydrogen. Hence, surveys for near-infrared jets are now an important means of detecting new Class 0 protostars. Long before an outflow has accumulated a detectable amount of cool CO gas, and before the protostar is detectable in the near-infrared or visible wave bands, the birth is already heralded by molecular jets. The lines are produced when the molecules are vibrationally agitated, requiring shock waves of at least 10 km s^{-1} to be present. With improving sensitivity, jets are now also being found in tracers of cooler molecular gas such as CO and SiO. It is also possible to detect cool atomic gas from some jets in the near-infrared but, in general, we would not expect to detect any optical jets due to the obscuration by surrounding dust.

Although there are examples of twin jets, molecular and atomic jets are usually found emanating from just one side of the source (see Fig. 10.2). Nevertheless, there are indications for the existence of a second jet in many of these outflows. Quite often, we are unable to detect either of the purported jets possibly due to the extinction. In such cases, we can attempt to disentangle a jet component from the region of impact where the jet is brought to a halt (§10.5). The jets rarely contain any smooth diffuse structure but more often consist of a chain of arc-shaped aligned clumps separated on scales of between 1000 AU and 10,000 AU. The clumps of emission are called 'knots' and are often bow shaped. Symmetrically located knots in twin-jet sources indicate that the knots originate from centrally-generated disturbances. These pulsations steepen up into shocks as they propagate along the jet. Within the pulses, compressed jet material accumulates to form dense cold bullets as detected in CO rotational lines. As the pulses move out along the jet, the strengths of the shocks decay and all the material is swept into the bullets. In this manner, a continuous jet is transformed into a chain of knots.

In the vicinity of the dust-enshrouded central stars, jets can still be found at radio wavelengths. The emission on scales under 100 AU testifies to the presence of aligned ionised gas jets, producing radiation through the free-free mechanism. In the radio, we can pinpoint the location of the source. With present continental-wide radio telescope networks such as the Very Long Baseline Array (VLBA), we can attain an angular resolution down to sub-milliarcseconds, which corresponds to sub-AU scale spatial scales in nearby star-forming regions. H_2O maser emission at 22.235 GHz is frequently detected in low-mass YSOs and is known to be a good tracer of jet activity very close to protostars while other masers are associated with protostellar disks. The masers are highly time variable on timescales of a day to a month. The maser luminosities correlate well with the luminosities of the radio continuum emission, also suggesting that masers trace the jet activity. They both appear to be excited in the shocked layer between the ambient protostellar core and jet.

Jets from Class I protostars are generally fainter. On large scales, these jets can often be detected in the optical emission lines. The main tracers are the Hα and [S II] lines (Fig. 10.3), which indicate that the jets are of low excitation typically produced behind shocks of speed 20–140 km s^{-1}. Molecular emission is often also observed coincident with the major optical atomic knots within the jets.

Several molecular jets display an almost linear increase in fluid speed with distance. Jet speeds inferred from proper motions of knots and radial speeds are generally quite low, between 40–100 km s^{-1}, in Class 0 jets. However, in the HH 111 jet, speeds above 200 km s^{-1} are found from both atomic and molecular components.

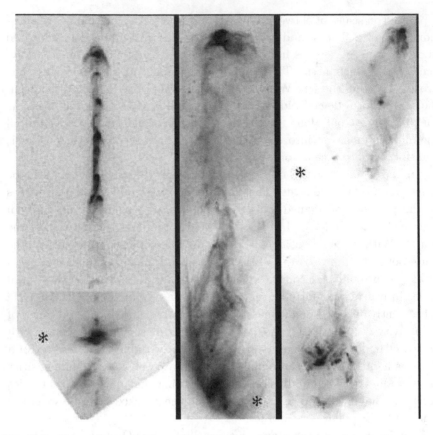

Fig. 10.2 The various optical manifestations of outflows driven by jets in atomic gas. The left panel displays the inner regions of the bright and knotty HH 111 (driving stars are indicated by an asterisk adjacent to their location). The middle panel displays HH 47, driven from a young star within the diffuse reflection nebula. A turbulent jet leads to a wide bow shock. Both these outflows possess counter-jets moving away from us (into the cloud) and, hence, show up better in near-infrared observations. The right panel displays the pair of HH Objects HH 1 (top) and HH 2 (bottom) driven by an obscured star close to the centre (not the visible star closer to the top). (Credit: NASA, HST & HH 111: B. Reipurth (CASA/U. Colorado) *et al.*, HH 47: J. Morse, NASA, HST. HH1/2: J. Hester (ASU))

10.4   Atomic Jets

The jets from Class II sources are almost exclusively atomic. Molecular signatures have disappeared. The line fluxes arise from within 1000 AU. After careful subtraction of the stellar continuum, microjet structure on scales under 100 AU can be revealed. In about 30% of the Class II sources, microjets are indeed found. These are short highly collimated optical jets with knots that display proper motions of a few hundred km s^{-1}.

Although these jets are weak and the accretion has subsided, the jets have attracted attention because they are observable at visible wavelengths. The Hubble Space Telescope (HST) and large telescopes with adaptive optics have given us glimpses of the structure on sub-arcsecond scales (§9.1). Through these observations, we hope to learn about the jet launching mechanism. Many strands of evidence have been uncovered from jets in the Taurus region, such as the following:

- The collimation occurs on scales large compared to the radius of the young star. Jets are only weakly collimated on the scale of $\sim$ 50 AU whereas the stellar radius is $\sim$ 0.02 AU.
- The jets are focused to within an angle of $\sim$ 10° after 1000 AU.
- The jets are knotty with the knots typically displaying proper motion of 100–200 km s^{-1}.
- The line profiles are broad and blue shifted to 100–400 km s^{-1}. The lower velocity is located around the edges and displays signs of acceleration with distance from the source.
- Jet speed tends to increase with the stellar luminosity.
- Variations in the jet speed are quite common.
- Asymmetries are observed in the radial velocity between the jet and counterjet.

In addition, the HST has helped to uncover irradiated jets where an external UV radiation field contributes to the excitation and visibility. These jets are produced by visible low-mass young stars in the immediate vicinity of OB stars. The young stars appear to have lost their parent cloud cores. The jets are either one-sided or with the brighter jet pointing away from the irradiating star and about an order of magnitude brighter than the counterjet. Spectroscopy shows that the fainter counterjets are moving several times faster than the main jets. Thus the brightness asymmetry reflects an underlying kinematic asymmetry. Furthermore, some of the Orion Nebula jets power chains of knots and bow shocks that can be traced out to 0.1 pc from their sources.

10.5   Herbig-Haro Flows

Besides the jets, a series of bow-shaped structures occur on larger scales. These extend out to several parsecs, forming the 'parsec-scale flows'. The apices of the bows face away from the protostar, giving the impression that they are being driven out (e.g. Fig. 10.3). Furthermore, proper motions are measured over timespans of a few years which demonstrate that the bow structure as a whole moves away from the source. In the reference frame of the bow, knots of emission move along the bow structure away from the apex into the flanks.

Such objects were first discovered on optical plates in the 1950s, and so obtained their name from their finders: Herbig-Haro (HH) objects. It soon became clear that the emission was excited by the particle collisions expected in shock waves. With higher sensitivity, we now often find complex structures which may extend back to the source. These are called Herbig-Haro flows. We also find outflow activity much further away from the source which suggest kinematic ages of around 30,000 years. Although this is still shorter than the Class I phase, it can be that the outflow direction has varied, so that the advancing edge we now observe has a much shorter kinematic age than the full duration of the outflow.

Models for HH objects as curved shock waves successfully reproduce the observed features. One identifying spectroscopic feature often found is the double-peaked line profile, signifying that the flanks of a bow shock deflect material both away and toward the observer. A problem encountered had been that the bow speed predicted by the spectrum is much lower than observed in the proper motion and radial velocity. How can the shock propagate rapidly without producing high excitation in the atoms? The only plausible answer was that the bow is advancing through material which has already a substantial motion away from the source. That is, many bows that we observe are not driven directly into the ambient cloud but are moving through the outflow itself. This implied that outflows were larger and older than previously imagined and that, if we looked hard and wide, we would find evidence for more distant HH objects. The discovery of these gigantic outflows was made possible by further developments in our instrumentation: we have developed detectors with the capability to image extremely wide fields, up to ten arcminutes wide.

Given the strong evidence that Herbig-Haro objects are bow shocks, they provide stringent tests for our understanding of shock physics and molecular dynamics. Some bows have been imaged in both molecular and atomic lines. It is then found that the molecular emission arises from the flanks of the bow, whereas the optical atomic emission originates from close to the apex. This separation is expected since the molecules do not survive the high temperatures near the apex. However, given bow speeds which typically exceed 80 km s^{-1}, the molecules should only survive in the far wings where the shock component has dropped below 24 km s^{-1}. In this case, the shock behaves hydrodynamically with two zones: a jump in temperature and density followed by a long cooling zone (J-shock). In a J-shock, essentially all the heating of the molecules occurs within a narrow zone, a few collisional mean-free-paths wide. Instead, we often measure H$_2$ emission from near the bow front. This factor and the low excitation of the molecules suggest that the shock surprise is being cushioned: ambipolar diffusion is softening the shock, spreading the front over a wide cool region which can be resolved. In this way, we gain the opportunity to understand how ambipolar diffusion operates within these C-shocks (where C stands for continuous).

If jets drive bipolar outflows, then we could expect to observe intense radiation from where the supersonic flow impacts against the external gas. While the external shock may correspond to a roughly paraboloidal bow, the jet is expected to be terminated by a round disk-shaped shock, which in fluid dynamics is called a Mach disk. Alternatively, if the impact is within the outflow itself then the interaction creates an 'internal working surface'. In this case, the two shocks are called a reflected shock and a transmitted shock. They sandwich a growing layer of outflowing gas accumulated through both shocks. In some outflows including HH 34, slower bow shocks are found at larger distances from the driving source, thus displaying an apparent deceleration. This suggests that the bow shocks lose momentum progressively when they drift into the external medium, once they are not directly driven by the jet.

The process by which large quantities of material are swept into the outflow is still not absolutely clear. Most evidence is in favour of a mechanism dubbed prompt entrainment in which the outflowing gas is swept into the wakes of extended bow shocks. The wings and wakes of the bows disturb the ambient material while travelling both within the outflow and advancing directly into the ambient cloud. Recent high resolution H$_2$ and millimetre interferometric observations provide supporting evidence. Alternatively, a wide slow stellar wind which surrounds the narrow fast jet would efficiently transfer momentum into the ambient medium. In addition, ambient core material may be entrained into the outflow across a turbulent

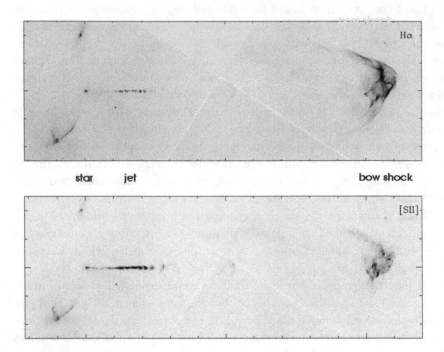

Fig. 10.3   The HH 34 jet and bow shock in two atomic emission lines. The bow shock is strong in the hydrogen tracer of warmer gas (top), while the jet only needs weak shock heating to appear in the sulphur line (bottom). (Credit: B. Reipurth *et al.*, AJ, 123, 362.)

boundary layer. The material is then gradually accelerated and mixed into the outflow.

10.6   Launch Theory

A scenario in which infalling gas can be partly ejected and partly accreted is desired. For this reason, understanding outflows may go a long way to determining how stars are made.

Three relevant classes of jet launching model appear in the literature. These are based on how the driving fuel is transferred into supersonic motion. The proposed driving forces are radiation, hydrodynamic and magnetic. The first two classes are usually dismissed on several grounds, especially since they are not all-purpose. Firstly, radiation pressure from the central young star lacks sufficient momentum to drive the molecular outflow. Photon momenta are roughly two orders of magnitude too low. Even in the high luminosity sources where most fuelling momentum, L_{bol}/c, is available, outflows cannot be driven. This could in principle be compensated for by multiple scattering of the photons (tens of times) within the outflow. Such an arrangement is not practical.

Hydrodynamic models can also be sub-divided into at least three types. The classical mechanism for smoothly accelerating a subsonic flow into a supersonic flow has been carried over from the theory of fluid dynamics. This involves a pressure gradient and a well-designed nozzle, called a de Laval nozzle. The astrofluid interpretation is the 'twin-exhaust' model in which two nozzles are located along the axis of least resistance. The nozzle shape and locations are determined by the pressure distribution in the protostellar core. The problems with this mechanism are that we do not possess the ambient pressure to contain the subsonic flow nor the solid nozzles which would resist break up as the fluid negotiates the transonic constrictions (i.e. the Kelvin-Helmholtz instability, §10.7). Alternatively, if we relax the requirement for a steady jet, we can obtain a high nozzle pressure by considering a rapidly expanding central cavity, replacing the ambient pressure by ram pressure and so construct expanding de Laval nozzles.

In the candle flame scenario, a central wind is deflected towards the disk rotation axis by an oblique shock surface. After deflection, the shocked fluid is cooled and compressed into a thin supersonic sheath which rams out of the core. In an elaboration on this theme, the central wind is more concentrated along the axis and a toroidal gas core is assumed. The resulting configuration generates wide bow-shaped shells which reproduce the shapes of classical bipolar outflows. These scenarios, besides not directly accounting for jets, do not produce bow shocks with the observed configurations.

In magnetohydrodynamic models, both the wind origin and collimation are accounted for. There are three potential launching sites: the disk, the disk-star interaction zone, and the stellar surface.

The disk wind model involves the gradual collimation of a centrifugally-driven wind from a magnetised accretion disk. The wind energy is derived from the gravitational energy released from the disk. This is accomplished via the gas rotation and a coupled magnetic field. The magnetic field lines behave like rigid wires, spun around as the disk rotates, and ejected packets of material behave like beads threaded on these wires. The crucial factor is the geometry of the magnetic field. If the magnetic field were aligned with the rotation axis, then angular momentum is transferred into the ambient gas through the propagation of torsional Alfvén waves and the disk is very gradually braked as already described in §7.6. On the other hand, if the field lines thread through the disk but are then bent by more than 30° from the spin axis, then the solid-body co-rotation throws out and accelerates disk material along the field lines. A nice feature of this model is that a small fraction of the accreting material is ejected, carrying with it a large fraction of the angular momentum. In other words, a disk wind permits accretion without the need for an anomalous viscosity as in standard disk theory (see §9.4.2). The main problem is to maintain the strongly bent magnetic field: an hourglass geometry representing squeezed field lines is required to preside after the core and disk contraction although the field lines have a natural tendency to straighten.

The MHD models for launching are complex and predictions difficult to extract. We can summarise some results here based on robust physical concepts. The angular momentum is carried by the rotation of the fluid and the twisted magnetic field. For a streamline originating from the disk at a radius R_o (i.e. the footpoint), the rotation like a solid body with angular speed Ω_o persists out to the Alfvén radius R_a where the radial speed of the escaping material has reached its Alfvén speed. Hence, within R_a Alfvén waves can propagate back towards the footpoints. Therefore, this distance corresponds to the lever arm which supplies a back torque onto the disk. The extracted specific angular momentum is $\Omega_o R_a^2$ and the terminal speed of a cold wind is

$$v_{wind} = 2^{1/2} \Omega_o R_a. \tag{10.1}$$

That is, the terminal wind speed is larger than the disk rotation speed by somewhat more than the ratio of the lever arm to the footpoint radius, which is estimated to be $R_r/R_o \sim 3$. To drive the accretion, the required mass outflow is

$$\dot{M}_{wind} \sim \left(\frac{R_o}{R_a}\right)^2 \dot{M}_{acc}. \tag{10.2}$$

Then, all the angular momentum of the disk is removed through the wind while removing $\sim 10\%$ of the disk mass.

Collimation is achieved through hoop stresses. At the Alfvén radius, the azimuthal field and poloidal field components are equal. Beyond this point, the azimuthal field becomes dominant and the Lorentz force bends the fluid towards the axis. The stability of the resulting jet is assisted by the jet's poloidal field which acts as a spinal column. Note also that the wind may be launched from a wide surface area of the disk and so the jet may thus display a strong shear with high speeds associated with the spine.

The alternative MHD model is called the X-wind model. As shown in Fig. 10.4, the wind is generated in the zone which connects the stellar magnetosphere to the inner region of a truncated conducting disk. This model, therefore, assumes that the disk does not extend to the protostar during the stages when jets are being produced. The inner disk, located where the disk co-rotates with the protostar or young star, provides the lever arm for launching the jets.

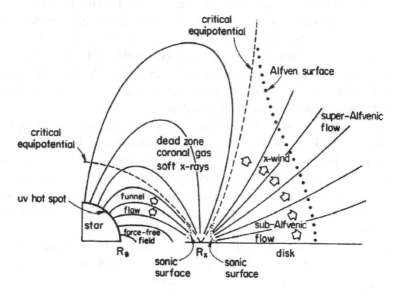

Fig. 10.4 The X-wind model for Classical T Tauri star environments – see text (Credit: F. Shu *et al.* 1994, Astrophysical J., 429,781).

Shielding currents prevent the threading of the disk by magnetospheric field lines. Therefore, the stellar magnetic field originating from the polar regions has to squeeze through the inner hole in the disk and is strongly compressed in the equatorial plane. With some magnetic diffusivity and in the presence of accretion, the field will manage to penetrate the innermost ring of the disk. This field threaded ring is termed the X-region.

The star has to adjust to the angular velocity of the inner disk edge in order to prevent a winding up of the field lines. Material in the innermost part of the X-region rotates at sub-Keplerian velocities (see §9.4.2) and is thus ready to move further in. The magnetic field channels this material in an accretion funnel towards regions close to the stellar poles, consistent with accretion theory (§9.4.3).

As the gas falls in, it would spin up due to angular momentum conservation if it were free. It is, however, attached to the rigidly rotating field lines and thus exerts a forward torque on the star and, more importantly, on the disk. The angular momentum of the accreting gas is thus stored in the X-region of the disk, which would be spun up. At the same time, the field lines threading the outer part of the X-region are inclined to the disk plane by only a very small angle (they have been squeezed through the disk in the equatorial plane from large distances). Hence, this part of the disk, rotating at super-Keplerian velocities, can launch a magneto-centrifugally driven disk wind: the X-wind. It is powerful enough to open the initially closed stellar field lines (which trace the weak field of the outermost parts of the stellar dipole), allowing the wind to expand. The result is that the X-wind efficiently removes angular momentum from the X-region which has been deposited there by the accretion flow.

The density as well as the velocity of the X-wind increase strongly but smoothly towards the polar axis. In the X-wind picture, the well collimated jets seen as Herbig-Haro or infrared jets are only the densest axial parts of a more extensive structure. The molecular outflow driven by the X-wind may thus be regarded as a hybrid of a jet driven outflow and a cavity swept out by a wide angle wind. In conclusion, the X-wind model is able to account for many observations in one fairly self-consistent model. The observations include time variable accretion/wind phenomena in T Tauri stars, the slow rotation rates of T Tauri stars, protostellar X-ray activity, and a number of the properties of jets and molecular outflows.

Jet launching mechanisms suggest that jets are not waste channels but important spin outlets. Without jets to remove the angular momentum, it is not clear what would then stop the young stars from reaching break-up speeds. Attempts are now being made to confirm that jets rotate. The measurements are difficult since angular momentum conservation predicts that the rotation speed falls as the jet expands. Therefore, the jet axial speed (and variations in it) tends to dominate the identifiable spectral features.

10.7   Jet Theory

We explore here the many reasons why jets are not smooth straight structures. The origin of the jet knots has been a subject of great debate. One initial idea was derived from laboratory experiments where gas jets of high pressure are known to expand and then converge on being injected from a nozzle. This produces oblique crossing shocks and, for specific geometries, the oblique shocks are intercepted by stronger transverse Mach disks on the axis. The result is the appearance of fixed shocks, which could be observable as a pattern of knots with a spacing of a few jet radii. Such a stationary pattern of knots has not as yet been found. In contrast, jet knots move at high speed (see §10.3 & 10.4).

A second knot production scenario which does produce moving knots is analogous to the mechanism which generates ocean waves. An interface between two moving fluids or gases is subject to the Kelvin-Helmholtz instability. The cause is the centrifugal force created when material slides over small corrugations on both sides of the interface. The effect is that small disturbances grow exponentially while they are advected down the jet. The advection speed is low when the jet is lighter than the surroundings which suggests that the instability may be significant in the Class II atomic jets rather than the heavy Class 0 molecular jets.

In terms of possible jet geometries – cylindrical or narrow cones – the instability produces coherent patterns involving pinching modes and helical modes. The penetration and bow shock formation are limited since the jet disintegrates once large-scale modes have developed. This is expected to disrupt a hydrodynamic jet on a length scale of 10–20 $M_j R_j$ where $M_j = v_j/c_j$ is the Mach number of the jet of radius R_j. On the other hand, other effects such as certain magnetic field configurations or a decreasing external pressure help provide for stabilisation.

A third cause relates directly to how the jet is launched. Variations in the jet speed, termed pulsations, will produce internal working surfaces. Even gradual sinusoidal variations will steepen into shock waves as the wave propagates down the jet. Given the knot spacing, ΔD_k, and the jet speed v_j, we estimate the duration between pulses as $t_k = \Delta D_k/v_j$ which can be written

$$t_k = 1000 \left(\frac{\Delta D_k}{2000 \text{ AU}} \right) \left(\frac{v_j}{100 \text{ km s}^{-1}} \right)^{-1} \text{ yr.} \qquad (10.3)$$

This implies that enhanced ejection associated with enhanced accretion provides a natural explanation for jet knots although it is not obvious how velocity variations, rather than jet density variations, result.

There are indications in many outflows that the jet direction is not fixed. There are wiggles in jets, meandering Herbig-Haro flows, widely spread Herbig-Haro objects and wide-angle outflows. The time scale for these changes can be just tens of years or up to 10,000 years. The change in angle is generally less than 10° although there are examples where it reaches 40°. The cause of the changes in direction is not known. Suggested causes are as follows:

1. The jets' direction follows the spin axis of the accretion disk and the disk is forced to precess by the tidal force of a companion star. Globally, the bipolar flow possesses a point or 'S-symmetry'. The precession time scale is thought to be about 20 times the Keplerian rotation period of the inner collimating disk i.e. it could be a few hundred years. A badly misaligned disk would re-align with the binary orbit in about a precession period due to energy dissipation from tidal shearing.

2. The disk mass is replenished on a timescale of $M_d/\dot{M}_d$ in the protostellar phases. Or, a merger event between the disk and another condensation may occur. The disk may thus consist of matter with a changing direction of mean spin. Although it depends on the particular source and stage of evolution, the time scale for disk and jet re-orientation probably exceeds 1000 yr.

3. The jet direction into the surrounding cloud is composed of two components: the jet velocity relative to the protostar and the motion of the protostar relative to the cloud. The jet direction is given from the vector sum of these components. If the protostar is part of a very close binary, its velocity component may be comparable to the jet speed and, hence, it produces prominent wiggles in the jet as it orbits. More likely, wiggles of the order of 1° should be quite common. Wide precession would imply a short orbital timescale – of order of years. In addition, a global reflection or 'C-symmetry' is predicted.

4. The accretion disk is warped by radiation from the central star. If the central star provides the power rather than the accretion itself, then the disk is more likely to be unstable to warping. However, warping is much more likely in the outer disk, on scales which do not determine the jet direction.

Each possibility thus has its own peculiarities. Yet, however good our interpretation, we require confirmation which can only really come from resolving the features within the launch region.

10.8 Outflow Evolution

10.8.1 *The jet flow*

We now gathered together and interpret observations which indicate that evolutionary relationships exist between the protostar, the accretion and the outflow. Here, we quantify these relationships in order to see how a stellar system might be constructed.

The outflowing mass and energy can be estimated through various measurements of the strength of the jets, the outflow and the impact at Herbig-Haro objects. One general method is to constrain the density, temperature and degree of ionisation and so construct a physical model. A second method is to measure the total luminosity from a single emission line and extrapolate using plausible interpretations.

Class 0 molecular jets possess temperatures in the range of 50–200 K and mean densities in the range of 10^4 to 10^6 cm^{-3}. The high densities are deduced from the properties of CO bullets and intrinsic molecular shocks, which excite H_2. Therefore, molecular jets are hypersonic with Mach numbers exceeding 100. They are probably also super-Alfvénic although the magnetic field cannot be tightly and directly constrained. The mass outflow from twin jets of radius r_j is then $\dot{M}_{jet} = 2\pi r_j^2 \rho_j v_j$ where the mass density is ρ_j, assuming a circular cross-section for simplicity. We thus obtain high mass ejection rates:

$$\dot{M}_{jet} = 1.3 \times 10^{-5} \left(\frac{r_j}{500\,\text{AU}}\right)^2 \left(\frac{n_j}{10^5\,\text{cm}^{-3}}\right) \left(\frac{v_j}{100\,\text{km s}^{-1}}\right) \ \text{M}_\odot \ \text{yr}^{-1},$$

(10.4)

where the hydrogen nucleon density is n_j and the molecular number density is $0.5\,n_j$ (noting that molecular hydrogen is by far the most abundant molecule).

In contrast, Class II optical jets typically possess high speeds and low densities. The density of electrons is usually derived, and we then require knowledge of the ion level now thought to be only a few per cent. Then, the total density is found to be only $\sim 10^3$ cm^{-3} even on the smaller size scale of these atomic jets. The overall result is a much lower mass loss rate:

$$\dot{M}_{jet} = 1.6 \times 10^{-8} \left(\frac{r_j}{100\,\text{AU}}\right)^2 \left(\frac{n_j}{10^3\,\text{cm}^{-3}}\right) \left(\frac{v_j}{300\,\text{km s}^{-1}}\right) \ \text{M}_\odot \ \text{yr}^{-1}.$$

(10.5)

This indicates an evolution in which the mass ejection rate falls as the accretion rate falls.

Can the supply from the accretion disk supply meet the demands of both jets and the protostar? In §7.8, we found peak mass accretion rates of order 10^{-5} $M_\odot$ yr^{-1} in Class 0 objects. According to Eq. 9.14, the typical accretion rate for Class II objects is $\sim 3 \times 10^{-8}$ $M_\odot$ yr^{-1}. Therefore, very large mass fractions appear to be ejected.

Can these jets supply the bipolar outflows? To prepare an answer, we revise the above jet formulae to provide thrust (momentum flow rates) and powers which can be compared to those derived directly from the outflow observations. The thrust per jet can be written as

$$\dot{F}_{jet} = 5 \times 10^{-4} \left(\frac{\dot{M}_{jet}}{10^{-5} \ M_\odot \ \text{yr}^{-1}} \right) \left(\frac{v_j}{100 \ \text{km s}^{-1}} \right) M_\odot \ \text{km s}^{-1} \ \text{yr}^{-1} \quad (10.6)$$

and the double jet power is

$$\dot{L}_{jet} = 8.2 \left(\frac{\dot{M}_{jet}}{10^{-5} \ M_\odot \ \text{yr}^{-1}} \right) \left(\frac{v_j}{100 \ \text{km s}^{-1}} \right)^2 L_\odot. \quad (10.7)$$

Note that the predicted jet power from Class 0 protostars is extremely high and is typically of the order of 0.1–$0.5 L_{bol}$. We can also now understand how molecules exist in these jets. If the jet contains a standard dust abundance, H_2 would form very rapidly at the base of a dense jet. Moreover, molecules will be formed particularly efficiently in the denser knots. Even in the absence of dust, H_2 can re-form in the knots through the H^- route discussed in §2.4.3.

10.8.2 *The bipolar outflow*

Three quantities are needed to quantify the energetics of a bipolar outflow. These are the total mass M_o, half the linear size D_o and the median outflow velocity V_o. Then, the kinematic age D_o/V_o and energy $(1/2)M_o v_o^2/D_o$ yield the mechanical luminosity of an outflow, defined as $L_o = (1/2)M_o v_o^3/D_o$. Similarly, the outflow thrust is defined as $F_o = M_o v_o^2/D_o$. It is not obvious how these are related to the present state of the jets and the driving protostar since the mass is that accumulated over the outflow lifetime and the kinematic age may be much smaller than the true age if the jet is precessing.

The bipolar outflow parameters are estimated from the CO emission lines emitted from low rotational levels since these record the accumulating cold mass on assuming some abundance of CO. The results so far obtained

indicate that the thrust and mechanical luminosity increase with the bolometric luminosity of the protostar. The thrust is usually presented since
momentum is conserved whatever the interaction, whereas energy is radiated away. Deeply embedded protostars drive outflows with thrust

$$F_{CO} \sim 2 - 10 \times 10^{-5} \left(\frac{L_{bol}}{L_{\odot}} \right)^{0.62} M_{\odot} \ \text{km s}^{-1} \ \text{yr}^{-1} \qquad (10.8)$$

over a broad range of L_{bol} from 1 $L_{\odot}$ to 10^6 $L_{\odot}$. Class I outflows are
relatively weaker with a relationship

$$F_{CO} \sim 2 - 10 \times 10^{-6} \left(\frac{L_{bol}}{L_{\odot}} \right)^{0.90} M_{\odot} \ \text{km s}^{-1} \ \text{yr}^{-1}, \qquad (10.9)$$

measurable over L_{bol} from 0.1 $L_{\odot}$ to 10^2 $L_{\odot}$. Therefore, outflows certainly
evolve with protostellar Class but the accretion rate (and hence the final
stellar mass) is the most critical factor. This is also apparent in another
quite close correlation, between the outflow thrust and the envelope mass
(as measured from submillimetre dust emission):-

$$F_{CO} \sim 0.5 - 8 \times 10^{-4} \left(\frac{M_{env}}{M_{\odot}} \right)^{1.2} M_{\odot} \ \text{km s}^{-1} \ \text{yr}^{-1}. \qquad (10.10)$$

We conclude that while accretion declines rapidly, ejection and accretion
remain closely coupled as if there is a single mechanism responsible for all
protostellar jets. It appears that high-mass fractions are ejected although
the actual size of this fraction is difficult to estimate. One feasible scenario
takes this fraction to evolve from 0.3–0.5 for Class 0, to 0.1 for Class I and
Class II sources.

10.9 Impact Theory

Besides the uncertainties concerning the momentum transfer into the jet,
getting it out again is also problematic. The standard approach to impacts
is based on equating the thrust which reaches the termination shock of the
jet to the rate at which stationary external material receives the momentum.
The speed at which jet material reaches the termination shock is $v_j - V_o$.
Therefore, ram pressure balance yields

$$\rho_j A_j (v_j - V_o)^2 = \rho_c A_c V_o^2. \qquad (10.11)$$

where the effective areas of impact, A_j and A_c, in the jet and cloud may differ. We shall put $\eta = \rho_j/\rho_c$ as the ratio of jet density to cloud density.

For heavy or ballistic jets, $\eta > 1$ and $v_j \sim v_o$, which means that the termination shock simply brushes aside the cloud and only a small fraction of the jet momentum will reach the external medium. If ballistic jets are associated with Class 0 outflows, then it would not be easy to understand the high mechanical thrust and luminosity. Furthermore, the bow shock should be the brighter shock due to the high speed but the weaker jet shock may show up as a non-dissociative shock i.e. a molecular knot.

On the other hand, if the jet is light, then the momentum is efficiently transferred but most of the energy remains within the cavity created by the shocked jet material. To determine the most favourable transfer conditions we write the thrust and power transferred at the impact as

$$F_{impact} = \rho_c A_c V_o^2 = \left(1 - \frac{V_o}{V_j}\right)^2 F_{jet} \qquad (10.12)$$

and

$$L_{impact} = 0.5\rho_c A_c V_o^3 = \left(\frac{V_o}{V_j}\right)\left(1 - \frac{V_o}{V_j}\right)^2 L_{jet}. \qquad (10.13)$$

Then, it can be shown that the maximum power ratio is $L_{impact}/L_{jet} = 4/27$, occurring if $V_o = V_j/3$ (and $\eta = 1/4$ if the areas are equal).

The observations suggest that the momentum transfer is probably more efficient than standard theory predicts. Higher efficiency may be achieved through precession which will better distribute the momentum of a heavy jet. In addition, the outflow age may then exceed the kinematic age, relaxing the demand on momentum placed by the simple models.

10.10   Summary

Outflow phenomena contain a fossil record of the accretion process, and the further study of bipolar outflows is expected to shed light on the star formation process itself. Outflows are thought to remove excess angular momentum from the accretion disks and to regulate the stellar rotation so that stars continue to grow without spinning up to break-up speeds. Most attractive in these respects is the X-wind model which combines a disk wind, driven by magnetic and centrifugal forces, with disk accretion via mass transfer to a stellar magnetosphere. MHD models can extract the angular momentum along with 10–30% of the disk mass. Such high mass loss rates are necessary to explain the power of molecular jets and outflows.

The impact of the jets and outflows could be critical to the star formation process. By feeding turbulence back into the core, and raising the sound speed, the outflow might temporarily speed up the accretion. If the bipolar outflow disrupts the cloud, however, then the accretion will subsequently be cut off. Finally, jet impact onto neighbouring cores could also trigger collapse in them. In regions containing many cores, sequential star formation could result.

Chapter 11

Massive Stars

Massive stars are low in number but make a large contribution to the properties of galaxies. They are fundamental to the production of the heavy elements and to the energy balance in the interstellar medium. They attempt to regulate the rate of star formation on large scales through feedback via intense winds, radiation and, finally, through supernova explosions.

Most stars are born in the neighbourhood of a massive star. Therefore, as viewed by the low-mass stars, massive stars are the influential or interfering relatives. The life history of a star is thus not only determined by the conditions at birth ('nature') but also the interaction with the environment in the formative years ('nurture'). Massive stars can lead to ejection from the cluster through their gravity or to a stunted growth through their feedback.

Their own origin remains a mystery. Moreover, massive stars should not even exist according to basic theory. This is because stars above 8 $M_\odot$ should 'switch on' their nuclear hydrogen burning during the accretion phase. Thus, their radiation pressure halts or even reverses the infall. This leads to the paradox that the hot O and B stars should not exist.

Studies of how massive stars form are afflicted by confusion. The problem is observationally severe because massive-star formation occurs in distant, highly obscured regions. On top of this, they are born in groups or clusters which hinders the study of the individuals. They are also theoretically difficult to analyse because of the many processes that must be acting simultaneously. The convenient well-defined stages found for low-mass star formation are missing. Despite all these obstacles, we need to explain their existence.

11.1 Basic Characteristics

A massive star is a term used for any star which will reach or has reached a mass exceeding $10\,M_\odot$. They are distributed tightly within the Galactic plane, confined within a disk subtending an angle under $0.8°$ with the mid-plane. The properties of the emission lines used to diagnose the spectra lead to their classification as O and B stars, or as Wolf-Rayet stars (hot stars with massive obscuring winds). In addition, their intense ultraviolet radiation ionises their surroundings, thus creating classical H II regions.

Their luminosity is typically 10^4–10^6 $L_\odot$. Once on the main sequence, the available hydrogen nuclear energy is supplied over a timescale $\tau_N \sim 7 \times 10^9$ yr $(M/M_\odot)(L_\odot/L)$. Therefore, they have short lives, spanning just $\sim$ 2–20 million years.

The time it takes to form a massive star is contentious. Arguments based on extrapolating from the mass accretion rates of low-mass star formation of $\dot{M} \sim\leq 10^{-5}$ $M_\odot$ yr^{-1}, lead to formation times exceeding 10^6 yr, a significant fraction of the main-sequence lifetime of the star. However, their total pre-main sequence life must be short. Given the Kelvin-Helmholtz or thermal timescale τ_{KH} as the ratio of the thermal energy to the luminosity, and applying the virial theorem, we define a lifetime

$$t_{KH} = G\, M^2/(R\,L). \qquad (11.1)$$

This provides a measure of how quickly a star would collapse in the absence of a nuclear energy supply and takes the value 3×10^7 yr for the Sun but is of order of just 10^4 yr for an O star.

In the traditional scenario of star formation, a star is initiated as an isolated cloud of gas. This sphere collapses at constant mass with no accretion or outflow. The evolution timescale is then simply the Kelvin-Helmholtz timescale. There is now, however, ample evidence for high accretion which we will discuss in this chapter.

The important comparison is between the accretion and Kelvin-Helmholtz timescales. Up to some mass, t_{KH} exceeds the free-fall collapse time t_{ff}, given by Eq. 4.1. This means that the accretion will be completed before the protostar has contracted. The mass limit depends on the details but probably lies in the range 8–15 $M_\odot$. Therefore, low-mass and intermediate-mass stars go through a phase where accretion has stopped but the release of gravitational energy still mediates the collapse. Accretion finishes long before hydrogen nuclear burning takes over. From then on, the protostellar evolution proceeds at constant mass.

In contrast, the high luminosity of higher mass stars means that $t_{KH} <$ t_{ff}. In this case, the central protostar has finished its contraction and is burning hydrogen during the accretion itself. Hence, massive stars begin their hydrogen burning phase while still in their natal dense cores. For this reason, the application of theory relevant to low-mass star formation is highly dubious and, to proceed, we first need to get familiar with a new set of observations.

11.2 Compact H II Regions

We cannot directly observe the optical and ultraviolet radiation from recently-formed stars. However, a massive young star emits ultraviolet photons in the 'extreme UV' (EUV), also termed the Lyman continuum (§2.4.2). These photons are not observed since they are spent ionising their immediate surroundings. However, the resulting H II region generates strong electron free-free emission at radio wavelengths. This radiation escapes and so produces a bright radio beacon, signalling the location of a hot star.

If we can estimate the age of the various beacons, we can learn how massive stars evolve. Normal H II regions around established massive stars are of size in the range 1–30 pc. The compact regions are of size between 0.005 pc and 0.5 pc and the electron density is in the range 2×10^3– 3×10^5 cm^{-3}. Ultracompact H II regions (UC H II regions) are those regions with size near or under 0.01 pc and the electron density above 10^5 cm^{-3}. Their morphologies have attracted much speculation: about 20% are cometary (compact head plus diffuse tail), 16% core-halo, 4% shell-like, 43% spherical or unresolved, and 17% irregular or multiple-peaked structures. We first investigate if these UC H II regions belong to the youngest massive stars.

The classical theory of H II regions assumes a uniform ambient medium and a sudden turn-on of the ionising flux. The first stage is very short-lived: the gas is ionised from the centre outwards, being led by an ionisation front. This expansion is halted when the ionised region is so large that the number of recombinations within that volume is equal to the number of photo-ionisations. This yields a volume of $4/3 \pi R_S^3$ where R_S is called the Strömgren radius. The total number of recombinations in unit volume is proportional, of course, to the number of close encounters, $\alpha_B n_e^2$, where n_e is the electron density (roughly equal to the ion density) and $\alpha_B = 2.6 \times$

10^{-13} cm^3 s^{-1} is the appropriate recombination coefficient. This coefficient excludes captures directly to the ground level since they re-emit an ionising photon whereas recombinations into excited levels produce photons which can escape from the region or are absorbed by dust. Equating the ionisation and recombination rates results in a sphere of size

$$R_S = 0.032 \left(\frac{N_{UV}}{10^{49} \text{ s}^{-1}} \right)^{1/3} \left(\frac{n_e}{10^5 \text{ cm}^{-3}} \right)^{-2/3} \text{ pc.} \qquad (11.2)$$

The number of UV photons covers a wide range. For main sequence stars classified as B2 ($\sim$ 20,000 K), $N_{UV} \sim 4 \times 10^{44}$; an O8.5 star ($\sim$ 35,500 K), $N_{UV} \sim 2 \times 10^{48}$; and for an O4 star ($\sim$ 50,000 K), $N_{UV} \sim 9 \times 10^{49}$.

The duration of this first stage is equal to the time required for the star to increase its output in the EUV since the region response time to the changing flux is very short, just $1/(n_e \alpha_B)$. After this, the internal pressure of the heated gas will drive a shock wave through the neutral surroundings. The expansion speed will be of order of the sound speed in the ionised region. The expansion finally stalls when the internal pressure has fallen so that pressure equilibrium is reached. This should occur at a radius estimated to be $(2 \, T_e/T_o)^{2/3} \, R_S$ where T_e/T_o is the internal-external temperature ratio.

The quantity that radio astronomers derive from the measured radiation is called the emission measure, EM, $n_e^2 \times L$, where L is the length through the region of electron density n_e, since this is proportional to the observed flux. From this, they find that, for a wide range of UC and compact H II regions, there is a relationship of the form $n_e = 7.8 \times 10^2 L_o^{-1.2}$ where L_o is the size. This would appear to agree with Eq. 11.2 only if the more compact regions are, in general, excited by less luminous sources. Another trend is derived from radio recombination lines of hydrogen: whereas line widths in typical regions are 20–30 km s^{-1}, with large thermal contributions, the compacter regions display dominant systematic or turbulent motions of between 25–60 km s^{-1}.

The ultracompact H II regions have been the centre of attention for another reason: there are far too many of them to be consistent with the classical Strömgren theory. Their number in the Galaxy indicates an average age of 100,000 yr, a hundred times longer than that given by expansion at the sound speed of 10 km s^{-1} to 0.05 pc. This is the *lifetime paradox* which models must try to address. In other terms, we estimate from observations that about 0.04 O stars are born per year, whereas the rate implied by fast expansion would be 0.3 O stars per year. Hence, the simple expansion of an ionised region is not a viable interpretation.

11.3   Models for Massive Star Environments

The lifetime paradox has found several interesting solutions. The range of solutions allows us to simultaneously explain the variety of compact H II region morphologies. We find that we cannot isolate a simple mechanism which can cope with all the facts and this leads us to understand the underlying complexity of massive star formation.

Bow shock models are ideal for explaining the cometary shaped regions. Where an ionising star moves supersonically through a cloud, ram pressure replaces thermal pressure as the limiting factor in the direction of motion. This model can be distinguished by obvious velocity structure in the high-pressure head region. Seen approaching or receding, the bow may appear circular.

A *champagne flow* is established if the massive star forms while near the edge of a molecular cloud. This is illustrated in panel (b) of Fig. 11.1. A cavity forms until the ionising front breaks out of the cloud. The ionising gas then streams away down the direction of least resistance. This forms a blister-type H II region, with a strong ionisation front on the cloud interior and a fan-shaped ionised region towards the exterior. This model is distinguished by high outward speeds in the fan or tail of the structure.

Strong stellar winds are also expected early in the formation. It is not clear exactly when the wind is bipolar. The wind produces a limb-brightened morphology from a very early stage as deposited momentum drives an expanding shell which is exposed to the ionising radiation.

The *mass-loaded wind* model invokes a clumpy environment, which can be expected through turbulence and fragmentation in the original cloud. The wind becomes mass-loaded as it ablates material from the surfaces of the clumps. The ionising radiation photo-ionises the clumps, reducing the size and extending the life of the H II region. Recent observations have yielded some awesome examples of such regions, with structures being eroded away to leave pillars and nests where the clumps may be triggered into collapse and further star formation, as shown in Fig. 2.2 and illustrated in Fig. 11.1.

Other models apply to more compact regions. Denser or more turbulent cores than have been considered as typical would provide the material to delay the expansion of the ionisation front and the pressure to resist the

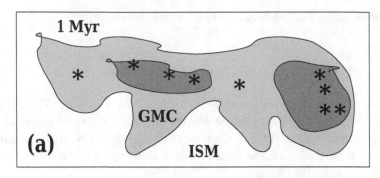

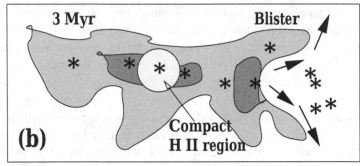

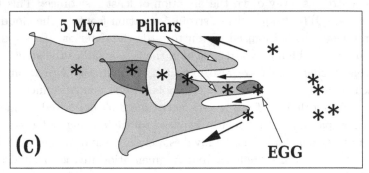

Fig. 11.1 A schematic diagram of the evolution of a high mass star formation region, illustrating a champagne flow, compact H II region, pillars and EGGs.

H II region. *Infall of gas* would prolong the compact state by providing more material and increased pressure. This may find application to some ultracompact and unresolved regions.

Photo-evaporation is expected to yield a dense wind from a disk. Moreover, massive disks are thought to surround massive stars although they have proven difficult to detect. The wind replenishes the gas in the H II region which is thus sustained for as long as the disk persists.

We conclude that massive stars are born in environments with a range of density, non-uniformity and disk masses and also with a range of dynamical states. Each star may develop along its own evolutionary path.

11.4   Hot Cores and Masers

A 'hot core' is a title bestowed upon a compact dense molecular core which is hot and dynamic. In the past few years, these cores have been shown to be the sites of massive star formation. Although most hot cores are intimately associated with compact and ultracompact H II regions, they are also found as their precursors. Hence, they are truly the 'cradles' of massive stars.

The hot cores possess temperatures in the range 100–200 K, density 10^6 to 10^8 cm^{-3}, masses from $100\,M_\odot$ to a few $\times 1000\,M_\odot$ and size 0.3–1 pc. The densest cores are traced in NH_3 and possess densities of 10^8 cm^{-3}, sizes down to 0.05 pc and temperatures of up to 250 K. These ammonia cores are often associated with very compact H II regions. The high temperatures lead to the evaporation of the icy mantles of grains, enriching the gas phase chemistry.

Hot cores are not isolated entities but are contained within Giant Molecular Clouds, within a hierarchical structure. Internally, the temperature and density decrease with distance from the heating star. Density gradients fall in the range $n \propto R^{-\alpha_n}$ with $\alpha_n \sim 1.5$–1.7 and $T \propto R^{-\alpha_T}$ with $\alpha_T \sim 0.4$–0.6. Line widths are high, 4–10 km s^{-1}, which provides evidence for a high accretion rate. There is also evidence for collapse, rotation and expansion in different cores.

The above properties suggest an early evolutionary sequence, beginning with a dense ammonia core. The core is under-luminous in the infrared and shows signs of collapse yet no UC H II region. An increasing luminosity is generated by an increasing mass accretion rate. A high accretion rate quenches the formation of an H II region. This situation is short-lived with radiation pressure onto dust grains reversing the infall. Then, an expansion phase begins which ends in the destruction of the core.

Masers appear in dense, active regions and so offer an opportunity to explore the changing conditions in the environments (see §2.3.1). They are proving useful for spotting regions of massive star formation. Water masers appear in small clusters of features which outline either disks or outflows. It is thought they appear during the earliest stages. Next, methanol masers are also signposts of disks and early protostars but often with detectable radio sources. Finally, OH masers appear to be associated with UC H II regions, possibly located in a shell of gas adjacent to a Strömgren sphere. Despite these assertions, our interpretations remain rather speculative. Nevertheless, masers sometimes provide indisputable evidence for jets and rotating disks on scales as small as 10 AU.

11.5   Outflows from Massive Stars

Molecular bipolar outflows are a basic component of all young massive protostars. Compared to their low mass counterparts, outflows from young massive protostar are much more energetic. Furthermore, massive molecular outflows are observed on large spatial scales (in the parsec range). Therefore, they are often easy to resolve spatially despite their typical distance of kiloparsecs.

The major question is: are these bipolar outflows driven by the same mechanism as those from low-mass stars? For example, the magneto-centrifugal wind scenarios predicts outflows which are morphologically similar to low-mass outflows and which show a high degree of collimation due to the star-disk interaction (§10.6). On the contrary, colliding protostars are expected to be extremely eruptive phenomena during which accretion disks should not be able to survive and, therefore, any resulting outflows are likely to be less collimated and rather appear more like explosions.

Many well-studied massive outflows have complex structures. A wide distribution of bow shocks, filaments and clumps are found. For example, the OMC-1 outflow in the Orion KL region consists of many bow shocks moving at hundreds of km s^{-1}. These bows may have formed as a dense shell of gas decelerated and fragmented. No evidence for a molecular jet is found although many show radio jets.

The studies agree on the basic facts that bipolar outflows from massive stars are ubiquitous and that they are very massive and energetic. However, there is disagreement on the typical collimation of the observed outflows. The average collimation of the classical massive outflows is lower than observed for their low-mass counterparts. However, in larger surveys and taking properly into account the poor spatial resolution, it is argued that the data are consistent with highly collimated outflows in high-mass star-forming regions. Single-dish millimetre observations are not sufficient to understand the complex bipolar outflows in massive star formation and proof will require interferometer observations.

The outflows are difficult to study because multiple outflows often emanate from the same large scale core. Clusters of stars form simultaneously in a core and the outflows originate from different protostars. For example, at least three molecular outflows are resolved in the core containing IRAS 05358+3543. This core contains an outflow with a collimation factor q_{coll} of 10, the highest so far associated with the formation of massive stars, and approaching the highest values recorded in low-mass star formation (§10.2).

The observations indicate that bipolar outflows in high-mass regions are made complex due to the strong UV radiation and stellar winds, in addition to the strong clustered mode of formation. Nevertheless, outflows of high collimation are produced by a physical mechanism comparable to that of their low mass counterparts. The protostars and outflows propagate within a crowded protostellar environment, which imprints very different structure onto the envelope and protostar. Thus, no other physical process has to be invoked, and high-mass star formation can proceed as in the classical low-mass scenario, with significantly enhanced accretion rates.

Much of the uncertainty and disparity in the results has been caused by attempts to unify all types of bipolar outlow. The wide outflows, classed by their low collimations and fragmented morphologies, may well be the result of the complex environment, combined outflows and stellar winds, and a poor hydrodynamic collimation mechanism, such as discussed in §10.6. It has been proposed that the models with accretion disks might not be appropriate for high-mass objects. Instead, massive stars may form through coalescence of stars rather than accretion. However, either scenario might lead to such wide outflows.

11.6   Accretion Theory

The physical processes controlling the birth of massive stars are still the subject of controversy. An accretion process is favoured for essentially two reasons: we observe disks and we observe massive bipolar outflows, both thought to be intimately linked to infall.

The following classical counter-argument shows that massive stars cannot form through direct radial infall. Radiation pressure on the gas and dust is able to reverse the infall, preventing the formation of stars with masses larger than 10 $M_\odot$. Explicitly, we require the gravitational acceleration to exceed the radiative acceleration of the dusty material i.e.

$$\frac{GM_*}{R^2} > \kappa \frac{L}{4\pi R^2 c} \tag{11.3}$$

where c is the speed of light and κ is the opacity. This equation yields a straightforward maximum opacity condition:

$$\kappa < 130 \left(\frac{M_*}{10 \ M_\odot} \right) \left(\frac{L_*}{1000 \ L_\odot} \right)^{-1} \text{cm}^2 \ \text{g}^{-1} \tag{11.4}$$

for inflow. Since the stellar luminosity increases as a high power of the stellar mass, this condition gets more difficult to satisfy for higher mass stars. Furthermore, dusty interstellar material has a very high opacity to the EUV radiation from hot stars, with $\kappa > 200 \ \text{cm}^2 \ \text{g}^{-1}$. In theory, we could overcome this problem by assuming that the dust opacity is significantly reduced, that the inflow is non-steady or that it takes place via optically thick blobs (such blobs could also be low-mass stars).

Inflow should, however, occur through an accretion disk. Therefore, radiation pressure is largely employed in blowing away the more tenuous spherical envelope, allowing material to sneak in through the massive disk. This is not clear cut, however, since the disk material must have originated from an extended but unprotected envelope. The radiation may also escape anisotropically, producing a 'flashlight' effect, occurring whenever a circumstellar disk is present. This permits material to approach the central source. However, at this point the material would again be rejected unless the dust has been largely destroyed.

Large and massive interstellar disks have been detected around massive stars. Rotation is suggested by the presence of smooth velocity gradients. However, circumstellar disks on scales under 10,000 AU have been difficult to detect although there is now plenty of good evidence mainly derived through maser detections. The EUV should photo-evaporate such disks on a timescale of 10^5 yr, forming an UC H II region.

However accretion proceeds, the standard star-forming theory predicts low accretion rates independent of mass (10^{-6}–10^{-5} $M_\odot$ yr^{-1}), just dependent on the sound speed (Eq. 8.5). At such rates, we would require over 10 Myr for a 100 $M_\odot$ star to form. Moreover, the inflowing momentum is not sufficient to overcome the radiation pressure of a star $\geq 8 \ M_\odot$. According to this scenario, more massive stars should not form although they are, of course, observed to exist.

The answer appears to emerge on assuming a much higher mass accretion rate. We now have envisaged the rate to exceed 10^{-3} M$_\odot$ yr^{-1}. In this case, the thrust of the infalling gas overcomes the outgoing radiation pressure. The cause of the high accretion rate follows from applying the fact that hot cores are supersonically turbulent. High turbulent velocities are indeed found in the hot cores (see §11.4) and, along with them, very high densities and pressures, conspiring to raise the expected accretion rate. The star formation time in this turbulent core model is several times the mean free-fall time of the core out of which the star forms but is about equal to that of the region in which the core is embedded. Thus, the high densities in regions of massive-star formation result in timescales for the formation of a massive star of about 10^5 yr.

In support, observations indicate that massive stars in the Galaxy form in regions of very high surface density corresponding to 1 g cm^{-2}. Clusters containing massive stars and globular clusters have a mean column density comparable to this i.e 10^3 stars pc^{-2}. The total pressure in clouds with such a column density is $P \sim 10^{-7}$–10^{-8} dyne cm^{-2}, far greater than that in the diffuse interstellar medium or the average in clumps and Giant Molecular Clouds given by Eq. 4.9.

The birthline concept introduced in §8.1 has been used as the basis to construct model evolutionary tracks on Hertzsprung-Russell diagrams. A single birthline is taken as the path followed by all protostars as their luminosity increases according to some fixed accretion rate formula (Fig. 9.2). For the typical case we consider, in which the cores have a density structure of the form $n \propto R^{-1.5}$, the mass accretion rate increases with time. The tracks do not divert from this line until the moment accretion halts, thus largely determining the final stellar mass. Therefore, in this theory, the birthline represents obscured objects while those off the line should be exposed, consistent with the observational definition of the birthline. If the accretion rate is low, then the birthline intercepts the main sequence once the accumulated stellar mass has reached ~ 8 M$_\odot$, implying there is no pre-main sequence stage for higher mass stars. Instead, they then proceed to move up the main sequence from the intercept. Therefore, massive stars may evolve along the main sequence provided they are still accreting.

Evolutionary models are fraught with uncertainty. We must still pin down how the infall of mass will develop and how much of this is ejected into the outflow. The infall rate also influences the deuterium available for burning. This, in turn, alters the protostar's radius, which is critical to the location on the Hertzsprung-Russell diagram.

11.7   Formation within Clusters

Complex patterns of radio continuum emission indicate that massive young stars are gregarious. They are born in associations and clusters. In addition, infrared surveys show that there are many more objects in the young clusters. Low mass stars to the tune of 10^3–10^4 stars per cubic parsec are found spread over regions of size 0.2–0.4 pc.

The crucial issue is how to accumulate sufficient material into a single star of mass 10–100 $M_\odot$. The gas accretion picture necessitates the presence of a massive turbulent core. However, we don't expect such cores to arise according to gravitational fragmentation. We have already calculated that the Jeans mass is typically just 1 solar mass in molecular clouds and will be only reduced within the dense environments within which massive stars form (§4.3.4).

One potential remedy is to delay the formation of massive stars until after the production of a large number of lower mass stars. This leaves warmer diffuser residual gas. The Jeans mass is then increased in the hotter cloud. Against this idea is that there is no observational evidence that the massive stars form later. Alternatively, we now understand how supersonic turbulence dominates the cloud dynamics. Consequently, we can envisage massive stars as forming directly out of massive cores generated by chance within a highly turbulent gas.

Segregation must also be explained: the most massive stars lie preferentially near the cluster centre. They cannot have dynamically evolved to the centre since the collisional relaxation timescale is longer than their ages. This has been explicitly shown for the Orion Nebula Cluster where the Trapezium stars are centrally located within a cluster of age ~ 1 Myr, too young to have dynamically relaxed. Furthermore, their distribution is not spherically symmetric, as would be expected if the 'swarm' of stars had relaxed. Therefore their location is a 'primordial' feature. That is, massive stars do not migrate to but are born in cluster cores. Note that, according to the Jeans criterion, the contrary should occur since high densities and high pressures favour the production of low mass fragments (see §4.3.4).

The merging of stars will produce higher mass stars and the higher collision probability in the crowded cluster core would naturally lead to the correct sense of segregation. The frequency of collisions depends on the effective cross-sectional area of the stars. While the actual cross-sections, πR_*^2 are tiny, the effective cross-sections will be given by the gravitational capture radius:

$$A_c = \pi R_c^2 = \pi R_*^2 \left(1 + \frac{GM_*}{2v_{disp}^2 R_*}\right) \tag{11.5}$$

where v_{disp} is the velocity dispersion of the stars. Adapting the molecular collision rate, Eq. 2.4, to stars, yields a collision timescale of roughly $t_C = 10^9/(n_s/10^4 \text{ pc}^{-3})$ yr on substituting typical parameters (e.g. $v_{disp} = 2$ km s^{-1}), which means that even with the high stellar density of $n_s = 10^4$ pc^{-3}, stellar collisions are far too infrequent.

There are theoretical possibilities to increase the collision rate. First, one can imagine that residual gas lying in the outskirts will move into the cluster core with the aid of tidal forces which disrupt incipient condensations. This has at least two positive effects. The residual gas can supply existing protostars, preferentially supplying the more massive stars as members of the population compete for the available gas. As well as this 'competitive accretion' (see §12.5.2), the increased core mass forces a contraction, enhancing the stellar density and the stellar collision rate.

Secondly, there is also a high density of cores and one can consider core coalescence to produce massive cores. If the cores contain low-mass stars, these stars would then merge. The outcome of a close encounter is uncertain although it is clear that the two stars form a binary with a decaying orbit. Simulations show, however, that such tight binaries disperse the remaining stars. Thirdly, and as a counter-argument, one can consider the protostars to be extended objects, possibly with massive disks. This increases their effective cross-sections and the massive disks can soak up the excess energy, alleviating the dispersal problem.

These ideas could be combined into a three stage scenario for the collisional evolution of a large cluster. In the first, the cores coalesce. In stage 2, the remaining gas settles towards the centre. In stage 3, star–star collisions occur in the core. There are, however, too many uncertainties and we save further discussion for our presentation of multiple star systems in general (§12.5).

Massive stars are not exclusively members of rich stellar clusters. Recently, isolated examples of massive stars have been discovered in nearby galaxies. Massive stars have been found in association with only very small groups of lower mass stars in the bulge of the M51 galaxy. There are also reports of apparently isolated, massive field stars in both the Large and Small Magellanic Clouds. Therefore, we need quite versatile theories, encompassing wide concepts if we are to reach a consistent understanding.

11.8   Intermediate Mass Stars

In 1960, George Herbig suggested that Ae and Be stars associated with nebulosity are pre-main-sequence stars of intermediate mass. They are either the analogues of T Tauri stars in the mass range between 2 and 8 $M_\odot$ or in their radiative phase of contraction onto the main sequence. Specifically, the selection criteria of these 'Herbig Ae/Be stars' were (1) the spectral type is A or earlier, (2) the spectra display emission lines, (3) they are located in an obscured region and (4) they illuminate a bright reflection nebula in the immediate vicinity.

The first criterion ensures that the stars lie in the desired mass range. The second and third ensure that the stars are young. The fourth excludes stars that are projected by chance onto dark clouds. The third and fourth together exclude extended objects which have detached envelopes produced by violent ejections (e.g. planetary nebulae and Wolf-Rayet stars). Applying these criteria, Herbig proposed a list of 26 stars.

Nowadays, the criteria are not so stringently applied. More relevant is the fact that all known Herbig Ae/Be stars possess an infrared excess due to thermal re-radiation indicating the presence of dust in the form of a circumstellar envelope or disk. In addressing the question as to whether accretion disks around high mass stars are present, the Herbig Ae/Be stars play a crucial role, since these objects are the only higher mass pre-main-sequence stars that are visible at optical and infrared wavelengths. Additional characteristics of Herbig Ae/Be objects include an anomalous extinction law and photometric variability. Furthermore, the spectral range has now been widened to include cooler stars up to F8, to bridge the gap between Herbig Ae/Be stars and the boundary for classical T Tauri stars.

Whether or not Herbig Ae/Be stars are embedded in accretion disks has not been fully established. It is now clear that there are large differences between the most massive stars in this group (with spectral types B5-B0, Be), and the less massive Ae ones. In Ae stars, we have strong evidence for the presence of circumstellar disks from millimetre interferometry and from direct images in the visual. This is not the case for Herbig Be stars. However, even for Herbig Ae stars there is great uncertainty surrounding the structure of the disks, with arguments for the existence of a dust component which is roughly spherically distributed. As a consequence, the spectral energy distributions of Ae stars have been interpreted as originating from extended spherical envelopes of low optical depth.

The excess emission of Herbig Ae stars has been a long-standing puzzle.

A large fraction of the stellar luminosity is re-radiated between $\sim 1.25\,\mu$m and $7\,\mu$m, with a peak at about $3\,\mu$m. The solution is now thought to involve the nature of the inner walls of the disk. For these stars, dust evaporation in the inner disk, where the gas component is optically thin to the stellar radiation, is expected if the mass accretion rate is low. The result ids the creation of a puffed-up inner wall of optically thick dust at the dust sublimation radius. This can account for the near-infrared characteristics of the SEDs and suggests that, by interpreting the details, we are approaching a more comprehensive understanding of star formation.

11.9   Summary

How long do massive stars take to form? Observations of hot molecular cores suggest a formation time of 10^5 yr. An analysis based on observations of protostellar outflows suggests a similar timescale: 3×10^5 yr. The small spread in ages of stars in the Orion Nebula Cluster, where there is no evidence that the higher mass stars have formed systematically later compared to the lower mass population, provides an upper limit of 1 Myr.

Do massive stars form from rapid gas accretion or coalescence? Discriminating between these possibilities is a challenging observational task for one main reason: massive stars form in rich clusters emitting copious amounts of ionising photons that profoundly alter the surrounding environment. This makes it very difficult to deduce the primordial configuration of the molecular cloud which represents the initial conditions for massive star formation.

The search for high-mass protostars in the act of collapsing has been another major quest during the last decade. Studies are hindered: representative statistics are not available, with surveys for massive protostars suffering from severe bias toward the brightest infrared sources in the Galaxy. Moreover, massive stars form in a clustered mode: it is nearly impossible to resolve the forming cluster with current telescopes. However, dust millimetre emission is optically thin and, therefore, scales with the masses of the protostellar envelopes. Large-scale dust continuum imaging of rich molecular complexes may prove crucial to constructing representative samples.

In conclusion, the formation, birth and early development of high mass stars are not distinct phases. Since high mass stars are never isolated, their formation is ultimately tied to that of star cluster formation. Hence, the next step is to investigate star formation in the cluster context.

Chapter 12

The Distributions

The story from conception, through birth and infancy, now approaches completion. We have formed objects with characteristics that we hope will develop into stars. They can now undergo the ultimate test – a direct comparison to the finished product: the adult star.

Population surveys yield the social groupings and associations. Although most stars are born in clusters, if they survive long enough they eventually gain their gravitational independency. These form the vast majority of the 'field stars'. While our Sun is now a field star, there are indications that it had a more interactive youth. Some others may have been raised in circumstances similar to their present relative isolation.

A population census yields the number of mature objects presently around us. In particular, we determine the fractions of massive giants and undersized dwarfs. Traditionally, we employ the star's mass as the single critical variable for the reason that this defines the main properties (luminosity, temperature and radius) once on the main sequence.

During the 1990s, we became adept at carrying out wide field surveys as well as probing into the close environment of the individuals. As a result, many promising scenarios have failed and had to be discarded. We find a high propensity of binaries and strong evidence which discriminates between their early binding and their pairing up in later life. We also find a certain propensity for segregation according to mass which helps differentiate between theories of mass migration and selective birth. Last but not least, we now detect brown dwarfs and free-floating planets.

12.1   Types and Prototypes

Any successful theory of star formation must be able to construct the following variety of objects. The ultimate fate of an object depends on its initial mass. We begin with the lowest conceived masses for gravitationally bound objects.

Planetary-sized objects (PSO). A PSO can form if a small mass $M <$ 0.01 $M_\odot$ is able to fragment and collapse. With this mass, no significant nuclear reactions, including deuterium burning, are triggered. The only internal source of energy is gravitational. The most accurate estimate now available is 0.011–0.013 $M_\odot$, i.e. 12–14 Jupiter masses. The name 'free-floating planet' implies that the PSO was once orbiting a star before being liberated. Whether they are escaped planets or independent collapsed entities remains open. The number of such objects is also difficult to estimate.

Brown dwarfs (BD). A BD is a sub-stellar object with a final mass in the range 0.012 $M_\odot$ $< M <$ 0.075 $M_\odot$. The central region gets sufficiently warm to ignite deuterium but this is not a durable fuel supply. Hydrogen fusion does not occur. Therefore, the object never reaches a state of equilibrium. Lithium is also likely to be burnt but after a considerable time (§9.1.2). Therefore, the presence of lithium confirms either an extreme youth or a brown dwarf status.

Lower Main Sequence Stars. These are hydrogen-burning stars with masses in the range 0.075 $M_\odot$ $< M <$ 0.25 $M_\odot$. Their interiors are fully convective but do not get hot enough to initiate the 'triple-alpha process' which fuses helium nuclei into carbon (nuclear physicists refer to helium nuclei as 'alpha particles'). These are also called Red Dwarf stars.

Solar-like stars are similar to Lower Main Sequence Stars with masses 0.25 $M_\odot$ $< M <$ 1.2 $M_\odot$ and their convective envelopes now contain radiative cores. Those with $M > 0.4$ $M_\odot$ will develop into giants. They also burn hydrogen through the proton-proton chain but will eventually become hot enough to fuse helium into carbon and possibly oxygen.

Upper Main Sequence stars possess mass $M > 1.2$ $M_\odot$. They possess convective cores surrounded by radiative envelopes. Core temperatures exceed 18 million K which means that hydrogen is transmuted into helium through a chain of reactions called the carbon cycle.

Giants & supergiants with $M > 4.0$ $M_\odot$ become hot enough to fuse carbon & oxygen into heavier elements (the central temperature reaches 600 million K) such as silicon, sulphur and iron. The iron nucleus is the end of the line as far as fusion is concerned, containing 26 protons in the most tightly bound of all nuclei.

Supergiant and Supernova progenitors possess mass $M > 8.0$ M$_\odot$. At these masses, iron fusion is then possible.

Wolf-Rayet stars are the short-lived descendants of O stars with $M > 20$ M$_\odot$. They have nearly reached the end of their stellar lives and an explosion is imminent. They are distinguished by strong obscuring winds from material being blown directly off the stars. Their spectral features are important diagnostics for estimating ages of clusters and starbursts.

The most massive stars that form within our Galaxy possess masses estimated to be of order 100 M$_\odot$ $< M <$ 200 M$_\odot$. Even more massive single stars are not excluded.

12.2 Binarity and Multiplicity

12.2.1 *The adult population*

A major advance in our knowledge has been the discovery that binarity is the rule rather than the exception. Most stars are members of binary or multiple systems. Estimates vary according to the type of star, with percentage membership ranging from 40–60%. Optical and infrared studies find similar percentages in nearby groups of T Tauri stars. Similar ages within a system suggest that the binaries form *in situ* rather than through accidental binding or later capture. Hence, binary systems are especially important to us because they contain a fossil record of fragmentation events which occurred quite early during star formation.

High resolution techniques used to explore stellar systems include speckle interferometry and adaptive optics. Both are based on the same idea: very short exposures contain much more fine detail of the sources than long exposures do because atmospheric turbulence cannot smear out the information. The trick is to process a large number of these short exposure frames using special algorithms. The term speckle interferometry was coined since stars really look speckled on the short exposures. Each speckle can be considered as an image of the source limited by diffraction.

We begin with the important details for adult stars. Our most reliable source of statistics is derived from the local population of low-mass mainsequence stars. The results are clear and unbiased.

1. Family membership. The frequency of occurrence for low mass or dwarf stellar systems can be expressed as single:binary:triple:quadruple sys-

tem ratios of 57:38:4:1, for mass ratios exceeding 0.1. For M dwarfs, ratios of 58:33:7:1 were found. However, there are hints for smaller companions and the estimate for the average number of companions that each primary star has is ~ 0.5–0.55.

2. Orbital periods. The periods are drawn from a wide distribution from less than a day to over 1 Myr. The distribution is single peaked and roughly log-normal i.e. $\log(P)$ rather than P is distributed like a Gaussian (§3.1). The median period, P, is 180 yr.

3. Separation. Multiple systems display a hierarchical structure (e.g. a distant companion orbiting a close binary) corresponding to long and short period sub-systems. The periods correspond to separations from a few $R_\odot$ to 10,000 AU with a broad peak at ~ 30 AU.

4. Mass ratio. We define $q = M_2/M_1 \leq 1$ as the mass ratio. Here, the data are not conclusive as yet. There may well be a peak in the distribution at $q \sim 0.23$ but the distribution below this value is not well determined. Brown dwarf companions may be relatively rare.

5. Eccentricity of orbits. Close binaries with periods of order days are in circular orbits. This could be expected from our knowledge of how tidal dissipation influences orbits during the main-sequence lifetimes of these stars and so does not provide clues as to the star formation processes. Wide binaries take on a range of eccentricities from ~ 0.1–0.9.

6. Age. Similar frequencies and statistics have been found in young clusters, Hyades (500 Myr old) and Pleiades (70 Myr old).

12.2.2 *The pre-main sequence population*

At the distances of the nearest regions of star formation, the lowest mass stars and brown dwarfs are very faint. We still have only hints about such properties as their mass accretion rates, rotational velocities and spectroscopic binarity (i.e radial velocity information). Binarity has been explored through diverse means: near-infrared speckle interferometry, lunar occultation and radio continuum observations.

1. Family Membership. Most stars form in multiple systems. For separations between 15 AU and 1800 AU, the binary frequency in T associations, such as Taurus and ρ Ophiuchus, appears to be about double that of the field. On the other hand, within the Orion Trapezium cluster the binary frequency is similar to that of the main-sequence.

2. The orbital parameters, periods and eccentricities, are quite similar

to those of main-sequence stars. The distribution of mass ratios of T Tauri stars in Taurus is comparatively flat for $M_2/M_1 \geq 0.2$, but the result is sensitive to the assumed evolutionary tracks. The mass ratio is neither correlated with the primary's mass or the components' separation.

3. Age. Binary formation has finished by the time young stars are a few million years old. The pairs in a binary appear to have the same age. Binary protostars have also been discovered but detailed information is still lacking.

12.3   Binarity: Theory

12.3.1   *Mechanisms*

We are not certain of the mechanism which produces binaries. We do know that the observations require that binary pairing and star formation proceed together. Potential explanations depend on the chosen combination of initial conditions and physical processes. We raise the following five possibilities here and show that the arguments narrow the field down to just one or two:

 (i) Capture: joining together of two unbound stars.
 (ii) Fission: bisection of one bound object.
(iii) Prompt Initial Fragmentation: early, during core formation.
(iv) Fragmentation during collapse.
 (v) Fragmentation within a disk.

Gravitational capture of point-like stars is extremely unlikely. Therefore, to enhance the direct capture argument, we search for processes which raise the probability of a close encounter which might lead to a capture. Removal of excess energy during an encounter can be caused by dissipation through the raising of tides on the two participating stars ('Tidal Capture') or by the transmission of energy to a third participating body ('Dynamical Capture'). However, both processes are found to be ineffective. Capture can be enhanced in dense clusters, or tighter sub-cluster locations, provided massive circumstellar disks are present to interact with ('Star-Disk Capture'). This favours small tight clusters where massive extended disks are more likely to survive the first encounter. This mechanism would naturally favour the formation of wide long-period binaries.

Fission scenarios require a contracting protostar to be unstable. If spinning rapidly, then a rotational instability could come into play as the contraction proceeds. Conserving angular momentum, the protostar will spin up and become unstable to axisymmetric perturbations when the ratio of rotational to gravitational energies surpasses a critical value calculated to be 0.27. The protostar deforms into a bar-shape and then the bar splits into two distinct bodies, forming a close binary. In this manner, spin angular momentum is converted into orbital angular momentum. This mechanism would clearly produce just close binaries. However, computer simulations have led us to abandon this attractive idea. The problem with it is that we are dealing with a compressible gas which will readily form spiral arms. The arms remove angular momentum, leading to a single central body surrounded by a spinning disk

12.3.2 *Fragmentation*

Gravitational fragmentation is one of the most popular themes for computer analysis. The complex physics, the extremes in scale and the variety of possible initial states make the numerical simulations challenging but the results often dubious. Nevertheless, the methods described in §5.5 have now reached some sophistication.

In the Prompt Initial Fragmentation scenario, we invoke an external agency such as a clump-clump collision. Clump collisions produce shock waves which help build up unstable dense layers capable of fragmenting into multiple protostars. For example, an initial mass of one Jeans mass, on a a scale of 10,000 AU, is disturbed and transformed into a compressed and distorted entity of several Jeans masses. The collapse then takes place simultaneously onto several gravitational centres. The gravitational centres are protostar-disk systems, which then evolve through dynamical interaction and energy dissipation to produce close binaries.

Fragmentation during collapse is the most versatile in theory. This can produce wide binaries, eccentric orbits, hierarchical clustering, disk and bar fragmentation. However, single stars are still predicted to result from the collapse of cores in which the density is centrally peaked. Cores in which the density peaks with $\rho \propto R^{-2}$ are quite resistant to fragmentation since each collapsing shell of gas has little influence on other shells. However, this now appears to be consistent with the latest findings as discussed in §8.3: pre-stellar cores are less compact and usually possess flatter density profiles.

Furthermore, pre-stellar cores are non-spherical (§6.5) and contain only a small fraction of their energy in their rotational motions (§6.2.2). This has prompted simulations of prolate and oblate clouds which are more prone to collapse into bars and rings, which subsequently fragment.

A second problem is raised by the standard scenario in which an inside-out collapse occurs. Here, one might expect that when about one Jeans mass of material has accumulated, it will promptly collapse directly to a single star. The scenario does not predict the formation of an object with at least two Jeans masses – the required mass in order to produce two separately bound fragments within a common envelope.

A third problem is to produce a sufficient number of close binaries with separations under 1 AU. Orbital decay is again invoked although other possibilities, such as very low angular momentum cores, remain to be explored. On the other hand, to see how to directly produce a binary on this scale, we use the formula for the Jeans length (Eq. 4.16) and $\lambda_J = 1$ AU to predict the required density: $\sim 10^{-10}$ g cm^{-3} or $n \sim 10^{14}$ cm^{-3}. This corresponds to a Jeans mass of < 0.01 M$_\odot$, on substituting into Eq. 4.19. In this case, fragmentation is only the beginning – it must be followed by massive accretion to build up the component masses.

Accretion of residual matter is expected to occur through a disk which will also influence binary. A companion will prevent a disk from forming in its vicinity, opening up gaps on either side of its orbit. It will also inhibit the expansion of a disk as it evolves on to larger scales, as calculated in §9.5. When the disk can no longer expand, the mass will empty out on quite a short timescale and the angular momentum can be transferred into the binary orbit through tides. Given the high percentage of binaries and the median separation of just 30 AU, it is feasible that half of the young stars will have their disks drained quite quickly, becoming relatively young (1 Myr) Weak-line T Tauri stars.

Interaction with a circumbinary disk would drive the system towards a tighter orbit. One can also argue that the accretion of gas with low angular momentum will lead to the merging of the components, whereas high angular momentum leads to a binary of near equal mass

Finally, disk fragmentation can be initiated by encounters or through disk instability as discussed in §9.4.2. In simulations, objects as massive as 0.01–0.1 M$_\odot$ have been found to form. Furthermore, early encounters may explain anomalous structure in our own solar system such as high eccentricities and inclinations.

These fragmentation models can be linked to the pattern emerging from observations of cores. First, we note that fragmentation is responsible for clustering as well as binarity and multiplicity. In principle, the distance between stars in binaries can take any value. There is, however, a pronounced 'knee' in the observed distribution at about a separation of 0.04 pc (8,250 AU). Apparently, this is the scale which divides clusters from binaries. This scale was found by Larson in 1995 and suggested to be correlated with the Jeans length. Larger systems form by fragmentation and independent collapse to form the independent envelope systems as mentioned in §8.3. A cloud containing many Jeans masses, often envisioned as prolate or filamentary, may thus form several independent cores akin to that expected through Prompt Initial Fragmentation.

In this picture, the common envelope systems (see §8.3) may be related to the fragmentation of cores more spherical and centrally condensed. In this case, fragmentation mechanisms work within the central region, especially when the density is not centrally peaked but quite flat, and proceeds according to the 'Fragmentation during Collapse' scenario.

Common disk systems, on the other hand, would appear to originate in clouds containing relatively high angular momentum. The disk forms early before splitting; tight stellar systems would result.

In summary, while a few plausible and some feasible binary mechanisms have been studied, direct observations of forming binaries may be necessary to really establish our knowledge. The wide range in periods and mass ratios, the number of close orbits, the range of eccentricities, the production of single stars are all in need of convincing explanations. Nevertheless, fragmentation (or the lack of it) at different stages during infall probably lies at the heart of the matter while dynamical events, rather than quasi-static collapse, greatly enhances the probability of fragmentation.

12.4   Nearby Clusters: Observations

Soon after their first exploration, Ambartsumian concluded that T Tauri stars were objects which had recently formed. The loose congregations of these youthful stars were termed *T associations*. They are unbound groups in dusty regions such as Taurus-Auriga, ρ Ophiuchus and Chamaeleon I. They typically produce a few thousand stars over a lifetime of 10 Myr. Because they are nearby, they provided most of our knowledge on the production of low-mass stars. It was assumed that low-mass stars were born almost exclusively in these groups.

It was also thought that high-mass stars were born in distinct clouds. Groups of these young stars were called *OB associations*, striking concentrations of short-lived brilliant stars. This has now proven false. The truth is that most T Tauri stars are also born in the environments of OB associations such as those found in Orion. Counting local stars, it is estimated that over 90% of the low mass stars younger than 10 Myr are born with OB association status. In addition, it has been proposed that our Sun probably formed in an OB association since that would automatically locate it close to a supernova, a possible origin of the chondrules found in the solar system (see §8.5).

It was infrared surveys which changed our view, enabling us to carry out a census of the stars still embedded in clouds. Infrared array cameras with large formats allow us to cover wide areas of molecular clouds. We now distinguish embedded clusters from exposed clusters. Rich clusters of stars (i.e. those containing over 100 stars) were identified within the clouds. We have discovered that star formation is not spread out evenly in a Giant Molecular Clouds but is strongly clustered. An example is the L 1630 cloud in Orion, where three clusters which cover 18% of the cloud contain over 96% of the stars. The three clusters are NGC 2071, NGC 2068 and NGC 2024. Ongoing star formation exterior to these regions is negligible.

Most stars form in clusters and so most stars interact while forming and should not be treated as isolated entities. Note that ρ Ophiuchus is also an example of the clustered mode with ~ 100 young stars within an area of $2\,\mathrm{pc}^2$. Clusters are, however, ubiquitous around young stars more massive than $\sim 5\,\mathrm{M}_\odot$. In contrast, the low-mass stars in Taurus are formed in an isolated mode with ~ 100 stars spread out over $300\,\mathrm{pc}^2$.

Stars clusters are centrally condensed. In young clusters we find that the surface density falls off as $\propto 1/R$ with distance R from the location where the number density peaks. Superimposed on this distribution are significant sub-clustering and structure. Cluster densities can reach several thousands per square parsec (e.g. Trapezium and NGC 2024) within small circular areas of radius 0.1 pc. Starless cores are of similar size (§6.2) and their mass density reaches values of $\sim 5 \times 10^3\,\mathrm{M}_\odot\,\mathrm{pc}^{-3}$. This is two orders of magnitude greater than presently found in optically-visible open clusters such as the Pleiades.

The spatial spread of young stars in such T associations yields information about past star formation. The typical dispersion in velocities of the stars is $v_{disp} \sim 1\text{--}2\ \mathrm{km\ s^{-1}}$. The lack of older stars gives rise to the so-called post T Tauri problem. This problem is that the older stars, which should have dispersed out of the cloud, are not found there. The velocity dispersion implies that these stars of age 10 Myr should be found in a halo of size 10–20 pc. Two types of solution have been offered: (1) Star formation in a cloud accelerates so that few of the older stars are to be expected in regions now industriously making protostars; (2) Star formation throughout a cloud is rapid. It begins and ends quite abruptly, lasting for not much longer than a crossing time of a shock wave, sound wave or magnetosonic wave. In other words, the dynamic timescale of an association and the free-fall timescale of an individual core are comparable.

The spatial spread of stars in OB associations is not always spherical. The youngest clusters, at least, maintain a record of the shape of the cloud from which they condensed. The Orion Nebular Cluster is apparently elongated with an aspect ratio (the ratio of major to minor axes) of 2. This probably stems from an even more elongated original cloud, estimated to have had an aspect ratio of 5. We explain this in terms of the process of 'violent relaxation', discussed below.

12.5   Cluster Formation: Theory

The mechanism by which a molecular cloud fragments to form several hundred to thousands of individual stars has remained elusive. The evolution from cloud to cluster involves three major steps: from gas to stars, dynamical relaxation, and cluster dissolution.

12.5.1   *From gas to stars*

The primary insight from large-scale computer simulations is that a stellar cluster forms through the hierarchical fragmentation of a turbulent molecular cloud. The supersonic turbulence manoeuvres the gas into sheets and filaments which occupy only a small fraction of the cloud volume. Fragmentation into many bodies also occurs most readily when the cloud contains filamentary structures.

The two observed modes of star formation are also reproduced in the simulated world. Isolated or distributed star formation occurs if the turbulence is driven. This inhibits wholesale collapse of a cloud. The driving scale and driving energy then determine the typical size of the star forming regions. Despite the turbulent cascade, the energy remains predominantly on the injected scales. Therefore, turbulent support breaks down on some smaller scales provided sufficient matter can be accumulated within a shock-compressed layer for gravitational forces to dominate locally.

The clustered mode occurs if the turbulence decays or is sufficiently intermittent so that large regions find no support over a dynamical timescale. A problem is to establish the initial state in which fragments are not forming on the small scale, yet turbulence is only driven on a large scale. In principle, one requires a high energy input on the small scales to prevent immediate collapse of isolated regions.

However, in a turbulent interstellar medium, an initial state with some form of quasi-static equilibrium rarely arises. When it does, the cloud is likely to be long-lived. Therefore, our observations of clouds, clumps and cores are all strongly biased towards these rare objects. Such a selection effect has almost certainly diverted our attention from dynamical models even though the turbulence was measured.

The most sophisticated SPH simulations are now able to follow the hydrodynamical evolution of quite rich clusters. First to develop are many small sub-clusters, which interact and merge to form the final stellar cluster. As opposed to a more uniform distribution arising from a monolithic formation, the hierarchical nature of the clustering implies that a protostar has a higher number of close neighbours and more frequent dynamical interactions. Such close encounters can truncate circumstellar discs, harden existing binaries and potentially liberate a population of planets. It is estimated that at least one-third of all stars, and most massive stars, suffer through such disruptive events.

12.5.2 *Cluster relaxation*

Groups of young stars can be rich or poor in number, bound or unbound gravitationally, and clustered or isolated in mode. They also tend to be segregated in mass, with the massive stars forming in the cluster centres (see §11.7).

Young clusters of stars are subject to two types of dynamical evolution. The first is a contradiction in terms: violent relaxation. This involves the global change of the cluster's gravitational potential due to collapse. The timescale is given by an average crossing time, $t_{cross} = 2\,R_c/v_{disp}$ where R_c is the radius of the cluster within which half the mass resides. The effect is independent of stellar mass since the gravitational potential of the entire cluster determines the dynamical evolution. It generates a centrally condensed configuration which slowly removes the record of the original state. However, the removal is slow and the imprint of an initial non-spherical configuration may be recognised.

The second type of evolution is called two-body relaxation (again, an apparent contradiction in terms). Two-body interactions transfer kinetic energy to the lower mass body, which has two major effects: a tendency towards equipartition of object energies and the sinking of more massive bodies towards the cluster centre. This requires the much longer time $t_{relax} \sim t_{cross}(N/8\ lnN)$, where N is the number of cluster members, but does lead to a complete loss of memory of the initial state. This process, while always in operation, is too slow to explain the observed segregation of massive stars in young clusters (see §11.7).

Numerical simulations demonstrate that the gas accretion and the dynamical interactions are simultaneous processes. Observations also show that young stellar clusters are still gas rich. At the outset, material not bound to a particular core provides frictional drag while the individual cores compete to attract the gas. Therefore, the final mass of the stars depends on the outcome of the *competitive accretion*. Gas falls preferentially into the deepest part of the core potential. Protostars which loiter in the centre thus accumulate most mass and so become massive central stars. Protostars in binary systems will also grow fast due to their location near the centre (where dynamical capture is more likely) and due to their extra attraction. As a result of the competition, inequalities in mass are promoted and a runaway might ensue. As a result of the dissipation, the cluster becomes tighter and binaries closer.

12.5.3   *Cluster dissolution*

The vast majority of stars are field stars: free to move within the Galactic potential. So, how do clusters dissolve? There are two strong possibilities: feedback from protostars and binary-controlled ejection.

All clusters develop out of massive and dense molecular cores. Their rapid dynamical evolution is the cause of their high gas content: most of the gas has not had time to accrete or disperse. If this gas is removed abruptly in the star formation process, blown away by winds, H II regions or outflows, then the cluster becomes unbound. Hence, exposed open clusters will be quite rare, explaining why the visible clusters that we do see cannot account for all field stars. Alternatively, if the gas is removed less dramatically over several crossing times, the cluster may adapt and survive with reduced numbers.

The second mechanism relies upon a central massive binary which prevails over the cluster, interacting and ejecting lesser bodies. The ruling binary takes up the total potential energy. The timescale depends strongly on the number of stars with a dissolution time of $t_{diss} \sim t_{cross}(N^2/100)$. Therefore, small clusters are depopulated rapidly, making them difficult to find before they have dissolved. In rich clusters, however, this mechanism is very inefficient. One then requires the above feedback mechanism, enhanced by the presence of highly interactive massive stars.

12.6 Brown Dwarves and Planets

Arguments have been made against the formation of sub-stellar objects. Therefore, the successful search for brown dwarfs represents considerable progress. The formation of sub-stellar objects is now an intimate part of the star formation story. Theories of star and planet formation predict that stars ($M > 0.075\,M_\odot$) form during the dynamical collapse of a cloud core, while planets ($M < 0.012\,M_\odot$) form via the accretional coagulation of material in a circumstellar disk. The intermediate mass at which the dominant formation mechanism switches from collapse to disk coagulation, however, is not known. Since brown dwarfs are objects with masses intermediate between those of stars and planets, they offer an important link between star and planet formation.

The formation of brown dwarfs and even of the lowest mass stars remains obscure because of the difficulty in determining their basic properties (e.g. mass accretion rates, rotational velocities, and spectroscopic binarity) at young ages. Accurately determining these properties generally requires high-resolution spectroscopy. At the distances of the nearest regions of star formation, however, the lowest mass stars and brown dwarfs are too faint to have been included in all but the most recent high spectral resolution surveys.

A brown dwarf forms when sufficient mass is not accreted to raise the central temperature high enough to spark hydrogen burning. This is a result of the Pauli exclusion principle of quantum mechanics which forbids electrons to occupy the same state. Therefore, the density reaches the level where the electrons in the core become degenerate and degeneracy pressure provides stiff resistance to collapse and heating. The result is that the cloud compression is inhibited and the core stops getting hotter. Such a cloud fails to become a star and is called a brown dwarf (since the term 'red dwarf' was already allotted to the class of lower main sequence stars).

A principle reason for examining brown dwarfs here is to complete a

census of all objects formed during star formation. They are often identified as objects excluded from being stars: stars possess luminosities and
temperatures of at least 10^{-4} $L_\odot$ and 1,800 K. The 'lithium test' is based
on the prediction that low mass stars burn their lithium within their first
50 Myr of life, whereas those brown dwarfs with masses below 0.064 $M_\odot$
maintain their lithium abundance forever. Since these stars are fully convective, the surface lithium reflects the value throughout the star. Another
method to detect brown dwarfs involves the search for methane (present if
$T < 1,500$ K), and, to distinguish brown dwarfs from planets, the detection
of deuterium signatures.

Planets begin to take shape at the end of the Class II stage (0.001 $M_\odot$
necessary). About 60% of T Tauri stars younger than 3 million years possess
dust disks, compared with only 10% of stars that are 10 million years old.
The implication is that the disappearance of disks in older stars is linked to
the appearance of (unseen) planets. In our solar system, the comets in the
Kuiper belt (100–200 AU) and the Oort Cloud (0.1 pc) are the residuals of
the disk.

Therefore, planet formation is a by-product of star formation. In detail,
we can envisage the following sequence. During the infall, material that
does not fall onto the star forms a disk. During the Class 0 protostellar
stage, the dust sinks to the midplane of the disk, forming a dense opaque
sheet. Within this sheet, the dust clumps together and agglomerates into
larger objects, called planetismals. During the Class I protostellar stage,
the planetismals collide and build up to Earth-size planets and the cores
of giant planets like Jupiter and Saturn. During this stage, proto-Jovian
cores rapidly form beyond ~ 5 AU (the 'snow line' beyond which water
freezes, making much more mass available in the solid phase). Within
this radius, the solids available to make the terrestrial planets are metal
oxides and silicates. Therefore, the inner solar system slowly forms rocky
planetesimals. However, the early planetismals and any other objects are
swept into the protostar, producing some of the accretion variability we
observe.

During the Class II stage, surviving bodies can accrete atmospheres.
Proto-Jovian cores (10–20 Earth masses) gravitationally accumulate huge
amounts of hydrogen and helium gas from the disk. Planetary embryos of
terrestrial and ice giants form 1000–10,000 km diameter bodies. Considerable orbital migration occurs in the disk because the disk actively accretes.
In the Class III stage, Jovian planet formation has ceased but the terrestri-
als and ice giants are still forming through accretion. Finally, the residual

gas and dust in the disk dissipate by being either blown away by stellar winds or accreted onto the planets.

12.7 The Masses of Stars

The birth rate for stars of each mass can be determined from the number of star of each mass that we find around us, given their age. In turn, their masses can be determined from their luminosities, given a few basic assumptions contained in the theory of stellar evolution. In this manner, 'the rate of star creation as a function of stellar mass' was derived from the luminosity function by Edwin Salpeter in 1955. This distribution is now described by the Initial Mass Function (IMF). Salpeter found that the number of stars, N_s, in a complete volume in the solar neighbourhood could be approximately represented by a declining power-law when divided into bins of equal $\log \mathcal{M}$:

$$\frac{dN_s}{d \log_{10} \mathcal{M}} \propto \mathcal{M}^{-1.35} \tag{12.1}$$

for stars in the mass range $0.4 < \mathcal{M} < 10$ where $\mathcal{M}$ is the mass in solar units. (It is sometimes easier to distribute the stars into bins of equal mass, in which case the function is one steeper: $dN_s/d\mathcal{M} \propto \mathcal{M}^{-2.35}$.) We are now able to calculate the IMF more accurately. However, considerable uncertainties remain, as illustrated in Fig. 12.1. Systematic variations of the IMF with star-forming conditions, the 'Rosetta Stone' for theorists, have yet to be confirmed.

For young contracting stars, there is also considerable uncertainty in the mass-luminosity relation. Alternatively, one can determine the spectroscopic class of the stars and use models to predict the masses. This is not so error prone as it might seem since model masses for pre-main-sequence evolution can be justified by calibrating against the masses of specific stars where dynamical measurements of mass are available (e.g. from the rotation of a disk). For young clusters, near-infrared surveys detect very low mass objects below $0.01 \, \mathrm{M_\odot}$.

An analytical fit based on a log-normal distribution, which is a normal (Gaussian) distribution with $\log \mathcal{M}$ as the variable, is often discussed. Such a distribution is favoured in circumstances where many independent processes are contributing in a multiplicative manner. However, it is now excluded since it does not predict a sufficient number of high-mass stars.

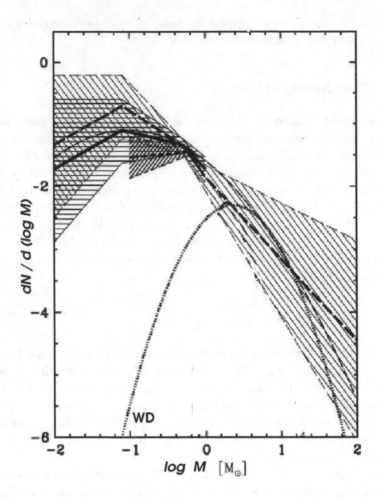

Fig. 12.1   The present-day IMF (thick-dashed) and Galactic-field IMF (thick solid line).
The errors are large: the shaded areas represent the approximate 95-99 per cent confidence region. A possible IMF for Galactic-halo white-dwarf progenitors is also shown
(labelled WD) and might indicate a variable IMF (Credit: from data originally presented
by P. Kroupa in MNRAS, 322, 231).

Instead, we fit a multi-component power-law which describes the flattened
distribution at low stellar masses and a further restriction or dearth in the
number of brown dwarfs. There is evidence that two breaks in the power

law are required:

$$\frac{dN_s}{d\log_{10}\mathcal{M}} = \begin{cases} 0.26 \ \mathcal{M}^{+0.7} & \text{for } 0.01 \leq \mathcal{M} < 0.08 \\ 0.035\mathcal{M}^{-0.3} & \text{for } 0.08 \leq \mathcal{M} < 0.5 \\ 0.019\mathcal{M}^{-1.3} & \text{for } 0.5 \ \leq \mathcal{M} \ . \end{cases} \qquad (12.2)$$

The indices are only indicative. There also appears to be considerable variation according to the precise population chosen. Nevertheless, a 'Universal IMF' can still be roughly adopted which has a characteristic mass of about one solar mass and a power-law tail at the high-mass end. Note that we can also trace where most of the mass is channelled: the mass function possesses indices one greater than the number function. This implies what we have already guessed: to order of magnitude, star-forming gas produces stars of one solar mass. In addition, very little mass goes into brown dwarfs and low-mass stars.

Important variations are beginning to emerge from cluster to cluster. The mass function in the Taurus star-forming region is quite narrow and sharply peaked. The young cluster IC 348 possesses more low-mass stars, giving a wider distribution. However, both IC 348 and Taurus are deficient in brown dwarfs (about 8% of the samples). In contrast, the Trapezium cluster and σ Orionis show no such deficit (over 20% are brown dwarfs). In fact, isolated planetary-mass objects may be as common as brown dwarfs, which together may be as numerous as the total number of stars.

To summarise, we emphasize that a reliable mass-age-luminosity relation is needed to derive an IMF. Assuming this, the galactic IMF is described by a power-law function for $M > 1$ $M_\odot$ and a log-normal form below. Star formation in the Galactic disk and in young clusters extends well below the hydrogen-burning limit and very likely below the deuterium-burning limit. The number of brown dwarfs is roughly equal to the number of stars, with a space density of order 0.1 pc^{-3}.

How do these facts tie in with the models? The near-uniformity of the IMF points toward a dominant self-similar, scale-free process. There have been a variety of explanations, some outdated, some updated and some still with merit. We list the following:

(i) A variety of mechanisms, involving many parameters, determine the mass of a star. Statistics then go according to the so-called central limit theorem (as if we combine values from a bank of random numbers).

This means that the number distribution should approach a log-normal distribution.

(ii) Mass regulation by protostellar feedback. The stellar mass might be determined when the outflow strength exceeds the inflow strength. Even here, the number of physical variables involved suggest that the number distribution might approach a log-normal distribution.

(iii) Accretion or competitive coagulation. The distribution is determined by collisions and mergers. A collisional hypothesis would seem less likely for the T associations.

(iv) An inherent property of supersonic turbulence. To verify, this requires numerical simulations involving a high number of Jeans masses and a wide range in scales.

(v) A reflection of the core mass function. The environment determines the distribution of gravitationally unstable cores which go on to form the protostars. The clump mass distribution is described by a substantially shallower power-law function than the IMF but the core mass function is perfectly consistent (§6.2.3). This was then interpreted to mean that the more massive *clumps* are less efficient in forming stars. They are indeed weakly bound or unbound and particularly prone to dispersal by their own protostars.

Given the many constraints on the speed of the star formation process, the turbulent production of cores, and the close resemblance of the IMF to the core mass function, it appears most straightforward to account for the distribution of stellar masses by supersonic turbulence in molecular clouds, preservation of the mass function during collapse followed by cluster evolution. However, this remains to be proven. According to the turbulence hypothesis, the higher frequency of brown dwarfs in richer clusters could be a result of their ejection from multiple systems before they have accreted up to stellar proportions. According to the fragmentation hypothesis, the higher frequency of brown dwarfs could result from the higher gas density in the star-forming environment, leading to a smaller minimum Jeans mass. Furthermore, a wider range in gas temperature would also broaden the mass distribution as inferred from the strong temperature dependence of the Jeans mass.

12.8 Summary

The ultimate data to test our concepts of star birth are the population census. Distributions in space and in mass can be measured for young populations as well as mature ones. Facts to highlight are the high number of binaries, mass segregation, a dearth of brown dwarfs in the low density environments so far explored, power-law high-mass tails and a characteristic mass (in the Galaxy field and Galaxy clusters) close to one solar mass.

A combination of hydrodynamics and stellar dynamics adequately explains most of these characteristics. However, the next few years of observations will certainly decide if other factors are essential.

Stars of each mass feature in a unique manner. The chemical enrichment of our Galaxy depends primarily on stars with mass exceeding $10\,M_\odot$ for their heavy-element content and the energy feedback produced by supernovae. The luminosity results mostly from stars with masses between 1 to a few $M_\odot$, and most of the mass is contained in objects with mass less than $1\,M_\odot$. Such mass distributions are found for galaxies in general as well as the components in the present day intergalactic medium (IGM). However, as we will now discover, it may not always have followed this pattern.

Chapter 13

Cosmological Star Formation

The first star in the Universe must have had remarkable properties. The first *generation* of stars must have been decisive in shaping the destiny of the Universe. So it is not surprising that the nature of these stars has fascinated astronomers. We have long known that they must have generated and widely distributed the metal elements. We now have learnt much more, culminating in a new burst of revelation following recent dramatic advances in our observational and theoretical capabilities.

It is now clear that star formation theory and cosmology are intimately associated. It seems likely that the first stars preceded galaxy and quasar formation. Direct observations of the environment which nurtured the first stars remain impossible since the events tool place while the Universe was in its 'dark ages'. Yet, many of the mysteries are no longer hidden and direct observation of that very first fiery light may not be too far away.

We now believe we can describe the scene. Observations have tied down the nature of the environment: the cosmological model. This means that in comparison to Galactic star formation, it is the perfect problem with simple well-established initial conditions: a flat universe composed of matter, baryons, and vacuum energy with a nearly scale-invariant spectrum of primordial fluctuations. Surely we possess all the information necessary to simulate the objects that will emerge?

There was a golden age when star formation was at its peak. The gas was trapped within collapsing and merging galaxies. Literally looking back in time, we witness a rise and fall in the star formation activity. Much of the formation occurred, as it still does, in gigantic bursts. These starbursts have been linked to the formation of quasars, elliptical galaxies and are probably also the sites of distant gamma-ray bursters, placing the theme of star formation at the hub of all astronomy. In this chapter, we explore the many variations on the theme as presented in different corners of our Universe. The problem is to unite this variety with a versatile theory of star formation. One possibility has emerged.

13.1   The First Stars

13.1.1   *The scene*

The equivalent of the Garden of Eden was a very pure environment for star profligation. The Universe was an un-enriched hydrogen and helium gas, with a mere touch of deuterium and lithium. There appears to be almost no cooling, radiation, magnetic field, turbulence or trigger to initiate or negotiate a collapse as occurs in the present state of the Universe. This all changed with the formation of the first stars. To understand how these stars appeared, we need to go back to the Big Bang.

The background is a spatially flat homogeneous Universe composed of ordinary matter, non-baryonic cold matter, radiation and dark energy. The cosmological framework is provided by the Cold Dark Matter (CDM) model for structure formation. This is now an established standard having survived many tests and alterations which make it exceedingly difficult to replace except with a model with essentially the same properties. Nevertheless, there are still inconsistencies. In the CDM model, structure grows from the bottom-up. In the present *Lambda*CDM version, a non-zero cosmological constant *Lambda* is included, which provides the large dark energy.

The Cosmic Microwave Background (CMB) was emitted when the Universe had an age of 380,000 yr. Like a blackbody surface, the radiation de-coupled from the matter at this point, and the ions and electrons recombined into neutral atoms. The temperature was then about 3,000 K. The expansion of the Universe has reduced this to the CMB temperature we now observe at 2.7 K in the millimetre waveband. The seeds for subsequent structure are detected as spatial variations in the CMB, produced from Gaussian fluctuations.

Something remarkable was recently reported which suggests that stars formed very early in the Universe. This breakthrough came from data acquired by the WMAP satellite. Launched in 2001, WMAP maintains a distant orbit about the second Lagrange Point, denoted 'L2', a million miles from Earth where it can accumulate data from the CMB with little light contamination. One of the biggest surprises revealed is that the first generation of stars to shine in the Universe ignited only 200 million years after the Big Bang, much earlier than many astronomers had calculated.

A new cosmological constitution has now been drafted. The WMAP team found that the Big Bang and inflation theories remain embodied. The expansion law for the local Universe is described by the Hubble constant, H, ($H = 71$ km s^{-1} Mpc^{-1}; i.e. galaxies separated by L megaparsecs recede from each other with speed $H \times L$). With this aid, we assemble the components. The Universe is composed of 4 per cent atoms (ordinary matter), 23 per cent of an unknown type of dark matter, and 73 per cent of a mysterious dark energy. The new measurements even shed light on the nature of the dark energy, which performs as a sort of anti-gravity. The dark matter interacts with ordinary matter through normal gravity.

What was particularly strange, however, was the timing of the first stars. This is ascertained indirectly through the influence these stars have on their local matter and the influence that this matter has on the passing CMB radiation. In the absence of these stars, the matter would have remained neutral from the time the CMB was emitted until the time the first quasars would have re-ionised the gas.

In order to comprehend time on these scales, the concept of redshift is central. The eras of star formation are described by the redshift, z. This is because we observe the redshifts to galaxies and quasars by identifying emission lines and their shift to longer wavelengths caused by the expansion of the Universe, i.e.

$$z = \frac{\Delta\lambda}{\lambda_{emitted}} = \frac{\lambda_{observed} - \lambda_{emitted}}{\lambda_{emitted}}, \tag{13.1}$$

where $\lambda_{emitted}$ is the true rest wavelength. Unfortunately, to convert this into an age has always required us to specify a model as well as the exact values of several parameters. Therefore, we have previously avoided this complexity by describing the evolution from the redshift of 1089±1 (the era of recombination) to the present redshift of zero. In terms of our standard, $z = 0$ corresponds to 13.7 Gyr (13.7 billion years), the age of the Universe with a margin of error of just one per cent.

The period of re-ionisation started with the ionising light from the first stars, and it ended when all the atoms in the intergalactic medium had been reionised. The most distant sources of light known at present are galaxies and quasars with redshifts z = 6–7, and their spectra indicate that the end of re-ionisation was occurring just at that time. However, WMAP demonstrates that there is considerably more electrons than this in the path of the CMB radiation: in order to explain this 'Thompson scattering', re-ionisation most probably started at redshifts 15–20. What is the origin of these stars?

13.1.2   *The first stellar nurseries*

The first stellar nurseries, the equivalent of Giant Molecular Clouds, were slowly being constructed during the dark ages. The dark ages began about 400,000 years after the Big Bang when the CMB photons shifted into the infrared, making the Universe completely dark to a human observer. The dark ages ended with the turning on of the first sources of light only 200 million years after the Big Bang. Therefore, between redshifts of 1000 and 15–20, the Universe was dark: no visible objects were in existence. Objects could not collapse because there were no coolants around to reduce the pressure and cause an implosion.

The cooling problem would finally be solved when a tiny fraction of molecules was able to form and persist. The only molecule which could form is, of course, H_2. However, H_2 in the interstellar medium forms on grains of dust and there is no dust in the Universe without stars to produce it. The alternative path is through reactions with electrons as described in §2.4.3. This is also problematic since there are also few electrons until there are ionising stars. Under these circumstances, primordial star formation relies upon the residual electrons from the recombination epoch or some produced in shock waves associated with collisions and mergers amongst small halos.

The story which emerges is a reconstruction of events based on numerical simulations and theory, constrained by indirect observations. Large gaps in our knowledge remain. The simulations demonstrate that the matter becomes increasingly confined to a network of filaments and material flow along these filaments. The massive halos, consisting of both dark matter and gas, are located at their intersections.

The conclusion is that the first star formed extremely early in the Universe. In fact, collapse could have taken place in very exceptional density peaks at redshifts $z \sim 40$. However, such a rare event will have little feedback effect on the Universe.

13.1.3 *The first generation*

The first generation of stars began to be produced in earnest after 180 Myr at $z \sim 20$. With the aid of gravity, the small background fluctuations grew into halos of dark matter. In the CDM bottom-up hierarchy, halos of increasing mass are assembled through mergers of smaller halos. At about a redshift of 20, the halos had reached a typical mass of $10^6 \, M_\odot$. The virial temperature in these mini-halos was below $10^4 \, \mathrm{K}$, which implies that they would be supported by thermal pressure until sufficient H_2 allowed the gas to cool.

We have a surprisingly good idea of the characteristic mass, M_f, for the first stars. This is because it is expected to be determined by fragmentation due to gravitational instability, i.e. the Jeans mass, which is determined by the gaseous state. Furthermore, there is a preferred gaseous state in the halos. The temperature of this state is $\sim 200 \, \mathrm{K}$, below which H_2 cooling is negligible since it is no longer rotationally excited (§2.4.5). The density of this state is $\sim 10^4 \, \mathrm{cm}^{-3}$ since collisions of H_2 with the dominant hydrogen atoms will dominate radiative cooling at higher densities (the so-called critical density, as discussed in §2.4.5). Therefore, the gas 'loiters' in this state allowing fragmentation to proceed.

We therefore adjust the Jeans mass from Eq. 4.17 into the form

$$M_J = 290 \left(\frac{T}{200 \, \mathrm{K}} \right)^{3/2} \left(\frac{n}{10^4 \, \mathrm{cm}^{-3}} \right)^{-1/2} \, \mathrm{M}_\odot, \qquad (13.2)$$

taking a neutral isothermal gas and n as the hydrogen nucleon density. Under the primordial conditions, we derive a mass of order of hundreds of solar masses for the first stars instead of the one solar mass predicted for cores in our Galaxy.

These stars would not resemble those found in our spiral arms and disk, called Population I stars. Population I stars are metal rich stars; like our Sun, they contain about 2–3 per cent metals and are just a few billion years old. Nor would the first stars be like those contained in the spherical component of the Galaxy (the halo and the bulge), called Population II. Population II stars are metal poor stars; they contain about 0.1 per cent metals and possess ages ranging from 2–13 billion years. Following, this sequence, the first stars, with no metals, are termed Population III.

Would the first stars have been so massive? It is very uncertain how much of the Jeans mass would be accumulated into a hydrogen-burning star. Processes associated with sub-fragmentation, accretion and radiation all raise the complexity. We expect a small core to form and accretion to proceed at a rate $\propto T^{3/2}$ (Eq. 8.5), perhaps a hundred times higher than for present-day star formation. There is also speculation concerning the formation of primordial brown dwarfs and low-mass stars. However, as we will see, there is considerable evidence that very massive stars do indeed form at high redshift.

13.1.4 *Early cosmological evolution*

Following the dark ages came the Middle Ages. The first stars immediately influenced their surroundings with both negative and positive feedback effects through their radiation and, on their death, through chemical enrichment. The pollution with heavy metals reduced the cloud temperature from 200 K (especially through oxygen and carbon fine structure line emission). Therefore, Population III stars were 'suicidal' and Population II stars were the offspring. The CMB temperature, falling according to the law $\sim 2.7\,(1 + z)$ K, maintains a lower bound to the gas temperature to provide a slowly-decreasing minimum Jeans mass.

The WMAP satellite has also revealed how the intervening gas alters the CMB radiation. First, the Universe was re-ionised with the birth of the first massive stars. However, with the quick death of the first stars through supernova explosions, the result was a partial recombination of the Universe. Apparently, the intergalactic medium was then not re-ionised until well after sufficient galaxy formation (e.g. beginning at $z \sim 10$ and complete at a redshift 6–7). This was then the Renaissance and our elaborate Universe had become fully visible.

Population III stars are extinct. Stellar archaeology is the science used to search for the relics of this population. For example, if we find many old low-mass low-metal stars in our Galaxy, then this would suggest that the Primordial Mass Function (i.e. the IMF of the Population III stars) is broad and extended down to low masses. At this moment, however, true Population III stars remain elusive and it is best to keep our opinions open until deep archaeological surveys are undertaken.

The so-called G-dwarf problem also testifies to a changing IMF. Although there is a modest increase of metallicity along with Galactic evolution from the spheroidal component (Population II) to the disk (Population I), there is a scarcity of very metal depleted stars in our Galaxy as well as in other galaxies: few low-mass stars were formed when the metallicity was very low at early times. A top-heavy IMF is also indicated for young galaxies at $z > 1$ through submillimetre and far-infrared surveys.

How can the theory be rigorously tested? The sources responsible for the high level of ionisation discovered by WMAP should be directly detectable out to $z \sim 15$ by the James Webb Space Telescope, due to be launched around 2011. In the meantime, it has been proposed that massive stars give rise to the brightest explosions in the universe: Gamma-Ray Bursts. Discovery of these bursts from very high redshift sources would throw open a new window into our medieval Universe.

13.2   Cosmological Star Formation History

There are many means of tracing star formation on galactic scales including UV emission and Hα emission. For the most active star-forming galaxies these signatures are hidden behind their own dust. Instead, the heated dust produces prodigious infrared fluxes which completely dominate the spectral energy distributions. Radio emission from supernovae is also a good indicator that star formation (involving massive stars) has taken place. On the other hand, confusion is caused by contamination from an old stellar population or an active nucleus (i.e. a quasar-like object at the galactic centre). To avoid gross errors, an amalgamation of observational techniques is now considered essential.

When were most stars born? It is now accepted that there was a golden age when the rate of star birth peaked. However, the historical facts are hotly debated. Even when the golden age occurred is questioned. The pristine gas which formed the primordial stars had been contaminated with metals. Star formation was no longer restricted to the massive fragments since the gas could now cool down. In the following, we express the formation rate in terms of the co-moving value, which adjusts the rate per unit volume by the expansion of the Universe.

When the Universe was 6.0 Gyr old, at a redshift of 1, the star formation rate was a factor of 10 higher than it is now. It had been thought that the cosmic star formation rate rose for the first 3.5 Gyr, until about the redshift of 2. The rate then fell dramatically as the fuelling gas became consumed. New data, however, suggest star formation peaked much earlier and with a very broad peak in the range $z \sim 3$–8. In this case, half of all stars had been formed when the Universe was just 3.2 Gyr old ($z = 2.14$) with only 25% forming at redshifts below 1.

It is evident that we occupy a Universe with a fast declining birth rate.

Nevertheless, low mass stars are resilient and the actual population continues to grow. This means that the average age of the population is rising: the average age is already 9 Gyr.

Where is the average star born? Data we have at present suggest that stars are born in galaxies built from quite massive halos. A wide range in sites with masses between 10^{10}–10^{14} $M_\odot$ seems probable. Our local group of galaxies, especially dwarf galaxies, should contain historical evidence. The dwarfs are the most common type of galaxy and could be the survivors of the first high redshift blocks which built massive galaxies. Individual stars can be distinguished since the Hubble Space Telescope came into operation. We can search for a fallow period of star formation caused by photo-ionisation following the re-ionisation. We find that the dwarf galaxies are indeed old but they also display a wide variety of evolutions. In contrast, distant galaxies are both difficult to detect and resolve. The radiation is redshifted and dust obscuration spoils our view. On the deepest exposures, HST 'Deep Fields' will resolve galaxies at redshifts beyond 6.

The first quasars are likely to have required more massive hosts at redshifts of 10 or higher. The most distant quasars now known lie at redshifts above 6. The space density of bright quasars, however, drops off very fast beyond $z \sim 2$ and they may thus not be sufficient to have re-ionise the Universe at $z \sim 6$, as required by observations.

13.3 Starbursts

A significant fraction of all stars are born in short episodic events. Such starbursts are unsustainable bursts of star formation; sufficient gas is not present for the formation of stars to continue for a time comparable to a galaxy's age. The duration of a starburst is typically a few tens of Megayears, rarely exceeding 100 Myr. Within this time, the star formation rate might exceed $100\,M_\odot\,\mathrm{yr}^{-1}$, thus consuming $10^{10}\,M_\odot$ of gas. Clearly, unless the gas is efficiently recycled, the burst must terminate.

We obtain our information on starbursts through all wavelengths. In particular, ultraviolet and infrared emissions testify to the massive stars being formed and the large supporting reservoir of gas and dust. Recent work has been prompted by the discovery with the IRAS satellite of infrared luminous galaxies. The extreme examples of starbursts are called luminous infrared galaxies (LIRGs) and ultra-luminous infrared galaxies (ULIRGs). The ULIRGS radiate over $10^{12}\,\mathrm{L}{-}\odot$ mainly in the infrared. The LIRGS radiate over $10^{11}\,L_\odot$ and are the most numerous objects in the Universe with such power.

Starburst activity can take place in any location, in disks, bulges and cores as well as in distant arms which appear to be the result of tidal interactions. They are usually confined to regions of size 1 kiloparsec. The most intense are located near the centre of the galaxies and are called nuclear starbursts. The rate of star formation per square kiloparsec of galactic area can reach $10^3\,M_\odot\,yr^{-1}$ but, for some unknown reason, not exceed this value. In comparison, the average rate in our Galaxy is 10^{-3}–$10^{-2}\,M_\odot\,yr^{-1}\,kpc^{-2}$

The cause of starbursts is known. Gravitational interactions between galaxies trigger star formation activity within. Strong interactions and galactic mergers can trigger the LIRGS. The ULIRGS are advanced mergers in which an enormous amount of gas has been disturbed and now funnels down towards the galactic nucleus. Therefore, ULIRGS may be closely related to the development of massive black holes. These black holes are believed to be the monster central engines of active galactic nuclei, fuelled by further gas infall. This relates starbursts to the jet-powered radio galaxies and quasars.

The Antennae Galaxies provide a spectacular example of a starburst. Shown in Fig. 13.1 is a ground-based telescopic view (left) which emphasises the long arcing insect-like 'antennae' of luminous matter flung from the scene of the galactic accident. Zooming in on the cosmic wreckage with the Hubble Space Telescope (right), we discover a thousand bright young clusters of stars including super star clusters (SSCs) with masses as high as a few $10^6\,M_\odot$. Clouds of gas and dust distort our view, especially around the two galactic nuclei giving them a dimmed and nebulous appearance.

The SSCs in the Antennae present clues to the star formation history. The youngest ones ($< 5\,Myr$) are located at the edge of the dust overlap region between the two interacting galaxies. The oldest ($\sim 500\,Myr$) are far from this region. They are the result of bursts of star formation triggered by the collision. The 500-Myr-old clusters are believed to have formed during the first encounter of the parent galaxies. The two galaxies separated following their first encounter, only to merge later and conceive the younger clusters.

Starbursts in dwarf galaxies are very common, both in the local Universe and beyond. Furthermore, we now recognise that a large fraction of the star formation in the nearby starbursts takes place in compact, high luminosity star clusters (i.e. 'super star clusters'), rather than in distributed regions.

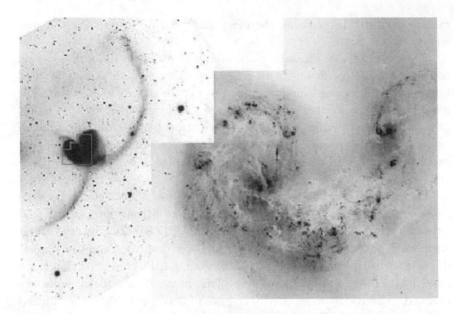

Fig. 13.1   The collision of the Antennae galaxies NGC 4038 and NGC 4039, 20 million parsecs away in the southern constellation of Corvus. The close-up HST optical image on the right, corresponding to the small inner area marked on the left image, has a pixel resolution of 5 parsecs. The image greyscale has been inverted to improve the display quality (Credit: Brad Whitmore, Space Telescope Science Institute and NASA).

There is an empirical law which relates star formation to the mass of gas in galaxies, displayed in Fig. 13.2. This is called the global star formation law or Schmidt law, after Maarten Schmidt who first assumed it in 1959. It confirms the importance of dense gas to star formation. Specifically, the rate of star formation per unit area, Σ_{SFR}, and the density of gas per unit area, Σ_{gas}, are found to be closely correlated in both normal galaxies and starbursts. A power-law of the form

$$\Sigma_{SFR} = A\Sigma_{gas}^n \qquad (13.3)$$

with $n = 1.4 \pm 0.1$ extends across a wide range of density. At low densities, however, the law is more complex, with a sharp decline in the star formation rate below a critical threshold density. The threshold density may be associated with the capability of forming massive clouds through large-scale gravitational or rotational instability of the galactic disks (see §9.4.2). Studies of this kind provide useful quick recipes for modelling star formation in computer simulations.

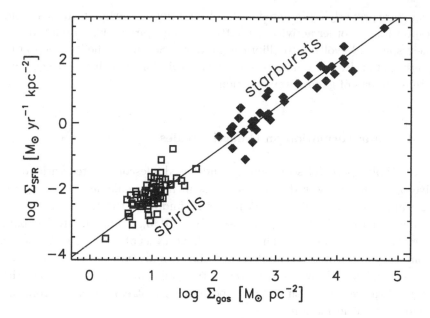

Fig. 13.2   The rate at which stars form in galaxies plotted against the measured amount of gas. Both are per unit surface area observed. The line represents the power law with $n = 1.4$. The rate is deduced from H_α emission, the gas from CO observations (data originally presented by Robert C. Kennicutt, Jr.in 1998).

In the local Universe, a quarter of all stars are born in starbursts. At $z = 1$, we estimate that the fraction is near to one-third with, however, only 4% of galaxies participating. Large-scale starbursts are thus very common features of early galaxy evolution. At high redshifts, the majority of the present-day 'normal' galaxy progenitors either appear to be undergoing violent gravitational interactions or experience very active star formation throughout.

This leads us to briefly digress from stellar origins to the consequences. A fascinating question is: what happens to a galaxy after a starburst? One idea is that ULIRGs evolve into quasars. Quasars are believed to be central supermassive black holes which are being activated by an abundant supply of fuel. Interactions and mergers between gas-rich spirals not only give rise to starbursts but also transport gas to the nuclear regions. Therefore, the central black hole should steadily grow in mass. In the final stages, the dust shrouding the black hole is blown away and an optical quasar appears. However, this scenario is controversial since not all the predicted stages are properly represented in the observations. Starbursts are also held responsible for other activities. Firstly, merging spiral galaxies tend to lose their spin and evolve into elliptical galaxies. Secondly, the production of massive stars leads to massive supernova explosions which might generate intense bursts of gamma ray emission.

13.4   Star Formation on Galactic Scales

The Hubble Space Telescope and ground-based telescopes fitted with adaptive optics have resolved stellar systems in nearby galaxies, revealing remarkable collections of stars. A recurrent finding is the existence of very young (typically 1–10 Myr), massive (10^3–$10^4 M_\odot$), and highly compact ($10^5 M_\odot pc^{-3}$) clusters. The clusters have evacuated the gas and dust from a region around themselves, forming a multitude of bubbles, shells and supershells. New stars are being triggered into formation along the edges of the regions, with pillars of molecular gas streaming away from the regions of star formation.

The 30 Doradus region is the prototype for these objects and cited as the Rosetta stone for understanding starburst regions. The region is located in the Large Magellanic Cloud (LMC) at a distance of 52 kpc. It was originally given a catalogue name appropriate for a star, but was soon recognized to be a nebula by Abbe Lacaille when he explored the still nearly virginal southern sky in 1751–52. The mistake was repeated in more recent times on a different scale, as divulged below.

The nearest giant extragalactic H II region has its home in 30 Doradus, consisting of expanding ionised shells, molecular gas and warm dust, concentrated in dense filaments which provide the apt description as the Tarantula Nebula. Fig. 13.3 also displays a very compact core called R 136, which is a super cluster containing some of the most massive stars that we know. Within R 136 lies several distinct components, each a rich cluster. R 136 was thought by some to be a single star of several thousand solar masses until speckle techniques, as explained in §12.2.1, resolved it into the brightest sources, which are now themselves revealed to be rich and compact clusters.

The region is regarded as undergoing intense enough star formation to be referred to as a mini-starburst. Star formation has occurred repeatedly over the past 20–25 Myr and is still continuing. The nebula itself is 200 pc in diameter within the region of size 1 kpc. There are over 1000 massive ionising O stars concentrated within a diameter of 40 pc with a compact

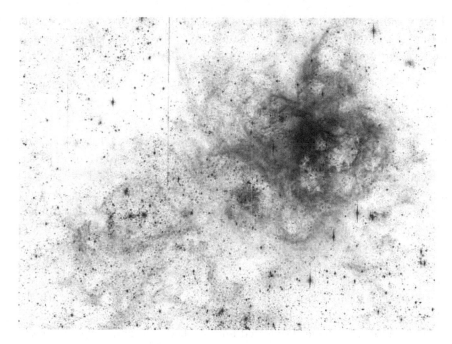

Fig. 13.3   30 Doradus is a vast star forming region in a nearby galaxy, the Large Magellanic Cloud. The optical image was taken with the HST Big Throughput Camera. The spindly arms of the Tarantula Nebula surround the NGC 2070 star cluster which contains some of the intrinsically brightest and most massive stars known (Credit: Gary Bernstein (U. Penn), Megan Novicki (U. Hawaii) & NASA).

core of massive stars, R136, 2.5 pc in size at its centre. Sub-clusters are found within, including R136a of size 0.25 pc. R 136 is a 'super star cluster' powered by ~ 100 massive stars. Interestingly, the latest burst of stars, occurring just one or two million years ago, appears to have temporarily suppressed the formation of lower mass stars. This may be connected with the richness since lower mass stars are identified in other poorer clusters such as the Eagle Nebula (M16).

The gaseous nebula shown in Fig. 13.3 is the result of the combined winds, ultraviolet radiation and supernova explosions from so many massive stars at a similar evolutionary epoch. We observe nested giant shells 20–300 pc in diameter that comprise the giant H II region. Surrounding 30 Doradus are even larger supergiant interstellar shells 600–1400 pc in diameter. These overlapping shells expand at ~ 50 km s^{-1}, while numerous $\sim$15 pc diameter regions exhibit outflows of speed 200 km s^{-1}. These are explained as young supernova remnants in the perimeters of giant shells.

The extended region does indeed contain a number of supernova remnants including the young supernova remnant 1987A, the nearest supernova that we have witnessed in recent years.

13.5 Globular Clusters

Globular clusters provide a particularly interesting test bed to investigate the Initial Mass Function (IMF) of stars (as well as the evolutions of galaxies and clusters of galaxies although that is not our topic). They provide a homogeneous sample of main sequence stars with the same age, chemical composition and reddening. Their distance is relatively well determined, allowing straightforward determinations of their luminosities. In addition, the binary fraction is small (under 20%). The major problem in determining the IMF of a globular cluster is to account for outside interference in addition to the internal relaxation processes pertaining to young clusters (see §12.5.2). A cluster gradually evaporates through interactions with the Galactic potential, interstellar clouds and other clusters along the orbit.

Fortunately, we have discovered and studied a large number of *young* globular clusters with ages typically 1–500 Myr. They are extremely massive ($10^3 - 10^8$ M$_\odot$), compact star clusters in merging, starbursting and even some barred and spiral galaxies. The brightest have all the attributes expected of proto-globular clusters, hence providing insight into the formation of globular clusters in the local Universe rather than trying to ascertain how they formed 13 Gyr ago. Nevertheless, only a fraction of them will maintain their compact status and become globular clusters.

The final field of star formation that we explore is the innermost 100 pc of our Galaxy. In fact, our Galactic centre harbours extraordinarily massive star clusters named the Arches, the Quintuplet and the Nuclear Young Cluster. The Arches is an exceedingly dense young cluster with a mass of around 7×10^4 M$_\odot$ within a radius of 0.23 pc to yield an average mass density of 1.6×10^6 M$_\odot$ pc^{-3} in stars. The total luminosity is $\sim 10^7$ L$_\odot$ generated by thousands of stars, including at least 160 O stars.

The Arches was formed in an abrupt, massive burst of star formation about 2.5±0.5 Myr ago out of a molecular cloud that has already been dispersed. The cluster is approaching and ionising the surface of a background molecular cloud, thus illuminating the thermal arched filaments. The Initial Mass Function (IMF) is again quite flat in comparison to the field star population (Eq. 12.1), similar to R 136, indicating a relative overabundance of high mass stars. The most massive stars are the mass-losing Wolf-Rayet stars with original masses of over 100 $M_\odot$. The cluster contains about 5% of all known Wolf-Rayet stars in the Galaxy.

Surprising is that a cluster could remain intact given the strong tidal shear exerted by the black hole at the centre of our Galaxy. Computer simulations have, however, demonstrated that some clusters could survive and be distinguishable until an age of $\sim$ 60 Myr, and several more may exist, slowly dissolving in the crowded region.

To summarise the Initial Mass Function results, there are indications for deviations from the general law given by Eq. 12.2. There may well be locations relatively deficient in low-mass stars and/or over-abundant in high-mass stars but the measurements provide a stiff challenge for our technology. Therefore, significant deviations still require confirmation.

13.6   Summary

Our understanding of cosmological star formation is emerging from the dark ages. The primordial objects were massive stars born when the Universe was just 200 million years old. These stars contaminated the Universe with metals, permitting stars, galaxies and quasars of all masses to form.

Galactic star formation history remains challenging. A theory which survives will have to be versatile. Yet, a paradigm is emerging. Galaxies are very individual but easily dominated, with gravity and turbulence as the two dictating processes. There are galaxies of low surface brightness within which the rate of star formation is low and neither dictator takes command.

Some galaxies are overwhelmed by a trigger: a merger. The merger instigates gravitational collapse which overcomes turbulence at first but eventually starts to feed it. Turbulence doesn't necessarily produce more stars, it just accelerates the conversion of some of the gas into stars and expels the rest.

Some galaxies are dominated by turbulence driven by an internal trigger: a shearing flow. The shear is generated from the galactic rotation. In this case, the star formation is widely distributed. A large-scale equilibrium rate is reached and the rate slowly decreases as the reservoir of gas is consumed. Other galaxies are overwhelmed by turbulence driven by another internal trigger: supernova and massive winds. This is a self-stimulating process, leading to a trail of star formation within a galaxy.

Some galaxies are overwhelmed by turbulence driven by an external trigger: a tidal encounter. The turbulence disturbs the gravitational field leading to gas infall and outflow, and gravitational collapse proceeds wherever the gas accumulates. This will be especially prominent in galactic nuclei where shear can be largely absent.

As has been discussed, the general turbo-gravity paradigm applies not only to galactic star formation but to star formation within molecular clouds. It is obviously not the entire story but provides the framework which renders preconceived initial conditions as unnecessary.

Epilogue

Past attempts have been made to assign the origin of stars to specific microscopic physical or chemical processes or to the establishment of specific initial states. The obvious attraction of possessing a law for star formation is clear. In contrast to this ideal, we now find ourselves involved with statistics and the macroscopic laws of astrophysical fluid dynamics.

We have found guiding principles. Stars originate through an interplay of the laws of physics and chemistry which steer the principles of turbulence and the nature of complex systems. In the simplest form, turbulence and gravity are the major contributors. For example, in some locations, supersonic turbulence gathers sufficient material together for gravity to take over. At other places, the gas is dispersed by the turbulence before it can collapse.

Yet during each stage of the story, other diverse processes could not be neglected. The six recent upheavals in our knowledge, described in the Introduction, testify to our need to better understand the behaviour of molecules, dust, radiation and magnetic fields within turbulent, rapidly evolving environments. In addition, the quest to uncover the origin of stars is replaced by the quest to find the origin of the turbulence.

We have discussed just one link in the chain of cosmic events leading from the birth of the Universe to the emergence of intelligent life. The story is still unfolding and we are well poised to make enormous advances in the early 21st Century. Although we believe we are making substantial headway, if we have learnt anything from the history of astronomy it is not to expect our new paradigm to remain unchallenged.

Index

Printed in the United States
By Bookmasters